PERGAMON INTERNATIONAL LIBRARY
of Science, Technology, Engineering and Social Studies
The 1000-volume original paperback library in aid of education,
industrial training and the enjoyment of leisure
Publisher: Robert Maxwell, M.C.

THE CONTROL OF GROWTH AND
DIFFERENTIATION IN PLANTS

OTHER TITLES OF INTEREST

BUCKETT: Introduction to Animal Husbandry

CLARK: The economics of Irrigation, 2nd Edition
  The value of Agricultural Land

CLAYTON: Agrarian Development in Peasant Economies

COWLING *et al.*: Resource Structure of Agriculture: An Economic Analysis

DILLON: The Analysis of Response in Crop and Livestock Production

DODSWORTH: Beef Production

DOUGLAS: Forest Recreation

FINDLAY: Timber Pests & Diseases

FUSSELL: Farming Techniques from Prehistoric to Modern Times

GARRETT: Soil Fungi and Soil Fertility

GILCHRIST SHIRLAW: Practical Course in Agricultural Chemistry

LOCKHART & WISEMAN: Introduction to Crop Husbandry, 2nd Edition

MILLER & MILLER: Successful Farm Management

NELSON: An Introduction to Feeding Farm Livestock

PRESTON & WILLIS: Intensive Beef Production, 2nd Edition

ROSE: Agricultural Physics

SHIPPEN & TURNER: Basic Farm Machinery, 2nd Edition

TAYLOR: The Role of Water in Agriculture

VOYSEY: Farm Studies

The terms of our inspection copy service apply to all
the above books. A complete catalogue of all books in
the Pergamon International Library is available on
request.

The Publisher will be pleased to receive suggestions
for revised editions and new titles.

THE CONTROL OF GROWTH AND DIFFERENTIATION IN PLANTS

by

P. F. WAREING, F.R.S.   AND   I. D. J. PHILLIPS

School of Biological Sciences,
University College of Wales,
Aberystwyth

Department of Biological Sciences
University of Exeter

PERGAMON PRESS

Oxford · New York · Toronto · Sydney · Paris · Braunschweig

Pergamon Press Offices:

U.K.   Pergamon Press Ltd., Headington Hill Hall, Oxford OX3 0BW,
                England

U.S.A.   Pergamon Press Inc., Maxwell House, Fairview Park,
                Elmsford, New York 10523, U.S.A.

CANADA   Pergamon of Canada Ltd., 207 Queen's Quay West,
                Toronto 1, Canada

AUSTRALIA   Pergamon Press (Aust.) Pty. Ltd., 19a Boundary Street,
                Rushcutters Bay, N.S.W. 2011, Australia

FRANCE   Pergamon Press SARL, 24 rue des Ecoles,
                75240 Paris, Cedex 05, France

WEST GERMANY Pergamon Press GmbH-3300 Braunschweig, Postfach 2923,
                Burgplatz 1, West Germany

First edition 1970
Reprinted 1971
Reprinted 1973
Reprinted 1975

Library of Congress Catalog Card No. 76-109055

Printed in Great Britain by A.Wheaton & Co., Exeter
ISBN 0 08 015500 6 (flexicover)
ISBN 0 08 015501 4 (hardcover)

To Helen and Anne

Contents

Preface

THE phenomenon of development, in both plants and animals, presents some of the most challenging unsolved problems of biology and is one of the remaining areas in which it cannot yet be said that we have made a decisive "breakthrough", comparable with the recent advances in biochemistry and molecular biology. Nevertheless, during the past 30 or 40 years there has been a steady advance in our understanding of the physiology of growth and differentiation in higher plants, so that there is now a considerable body of well-established knowledge in this field. It is generally accepted that some knowledge of the physiology of plant growth and differentiation is important for all students of the botanical sciences, and sections on this subject are commonly included in textbooks of plant physiology. The developmental approach is now also given more prominence in the contemporary teachings of plant form and structure than in the past, but there have been few attempts to bring together both morphological and physiological approaches within the same volume, and this we have attempted to do. Unless we attempt to relate the two approaches to each other, morphological and anatomical accounts of growth and differentiation must remain largely descriptive in nature, whereas our aim should clearly be to understand the processes underlying and controlling the structural changes. Conversely, physiological and biochemical studies which are not related back to developmental processes in the plant are liable to lose relevance and biological significance.

While it is true that we have found it convenient to concentrate on the structural aspects of development in the first few chapters, nevertheless we have attempted, throughout the book, to base our whole approach upon the growth and differentiation of organs and tissues, and the major developmental changes in the plant as a whole. Thus, after giving a brief account of the chemistry and biochemistry of plant hormones, we return to consider their role in the control of growth and differentiation at various levels. We then consider the factors regulating the major phase-switches in the development of the plant as a whole, viz. those controlling flowering, dormancy and senescence.

Finally, in the last chapter, we examine some basic problems in plant development, including the nature of the control mechanisms in differentiation. There are large gaps in our understanding of some of the most important aspects of development, such as the nature of the processes controlling the orderly sequence of changes so characteristic of development of all organisms. The discussion on such topics is, of necessity, largely speculative, but current thought in molecular genetics and theories of gene activation and repression derived from studies on micro-organisms enable us to formulate the problems in more precise terms than has been possible hitherto.

The book is intended primarily as an introduction to plant growth and differentiation for undergraduate students. While we have attempted, so far as possible, to give the supporting evidence for statements, references to the individual pieces of research upon which we have drawn are reduced to a minimum. We believe that an over-extensive chronicle of research workers' names and dates only serves to distract the student's attention from the main theme. We are confident that the many unnamed researchers whose work has made this book possible will not be offended. To assist the student, however, we have provided a list of additional reading, from which he can obtain detailed reference lists.

This book has only been made possible by the kind assistance of many persons in different ways and we make grateful acknowledgement of their help. We should mention specially, however, Professor A. W. Galston, Professor J. Heslop-Harrison, our editors, Professor G. F. Asprey and Dr. A. G. Lyon, and our colleagues, Dr. M. A. Hall and Dr. P. F. Saunders, all of whom have read sections of the book and have made most helpful suggestions. Responsibility for any errors which may have crept into the book must remain ours, however. We also wish to thank Miss M. Bigwood for her skilled assistance in the preparation of some of the diagrams.

CHAPTER 1

Growth in the Higher Plant

INTRODUCTION

We are all so familiar with the remarkable changes which occur during the life cycle of a plant, from germination to fruiting and senescence, that we tend to take the phenomenon of development for granted, so that it may cease to excite our wonder. Nevertheless, the orderly succession of changes leading from the simple structure of the embryo to the highly complex organization of the mature plant presents some of the most fascinating and challenging outstanding problems of biology. In this book we shall be concerned primarily with describing and examining what is known about the processes underlying and controlling plant development.

Before we can proceed, however, it is necessary to define certain terms, which are not always used in precise senses. *Development* is applied in its broadest sense to the whole series of changes which an organism goes through during its life cycle, but it may equally be applied to individual organs, to tissues or even to cells. Development is most clearly manifest in changes in the form of an organism, as when it changes from a vegetative to a flowering condition. Similarly, we may speak of the development of a leaf, from a simple primordium to a complex, mature organ.

Plant development involves both *growth* and *differentiation*. The term growth is applied to *quantitative* changes occurring during development and it may be defined as an irreversible change in the size of a cell, organ or whole organism. The external form of an organ is primarily the result of *differential growth* along certain axes. However, during development there appear not only quantitative differences in the numbers and arrangement of cells within different organs, but also *qualitative* differences between cells, tissues and organs, to which the term *differentiation* is applied. Differentiation at the cell and tissue level is well known and is the primary object of study in plant anatomy. However, we may also speak of differentiation of the plant body into shoot and root. Similarly, the change from the vegetative to the reproductive phase may also be regarded

1

as another example of differentiation. We shall, therefore, apply the term differentiation in a very broad sense to any situation in which meristematic cells give rise to two or more types of cell, tissue or organ which are qualitatively different from each other.

Thus, we may say that *growth and differentiation are the two major developmental processes*. Usually growth and differentiation take place concurrently during development, but under certain conditions we may obtain growth without differentiation, as in the growth of a mass of callus cells (Chapter 8).

The problems of development can be studied in a number of different ways, but basically there are two major types of approach, viz. (1) the morphological and (2) the physiological and biochemical. Developmental morphology and anatomy were formerly largely concerned with describing the visible changes occurring during development, but current interest is mainly directed to trying to understand the factors and processes determining plant form, using experimental techniques, such as surgery, tissue culture, autoradiography and so on. However, development cannot be fully understood without a study of the manifold biochemical and physiological processes underlying and determining the morphological changes, and it is these latter aspects of development which form the main subject of this work.

The experimental morphologist often uses the term *morphogenesis*, which, in the literal sense, is concerned with the origin of form in living organisms. However, by the term "form" should be understood not only the gross external morphology of the plant, but its whole organization, which may be recognized as existing at several different levels; thus, we may recognize (1) the structural organization of the individual cell, as shown by electron microscopy, (2) the organization of cells to form tissues, and (3) the organization of the plant body at the macroscopic level. Moreover, in the study of morphogenesis we are concerned not only with observable changes in form and structure but also with the underlying processes controlling the development of organs and tissues, and insofar as these processes must ultimately be explainable in terms of physics and chemistry, this aspect of morphogenesis is identical with developmental physiology and biochemistry. However, at the present time our knowledge of the molecular basis of morphogenesis is very fragmentary and we know very little about the physiological and biochemical processes regulating, for example, the initiation and development of leaves.

When we come to consider the physiology of development, we find a further dichotomy of approach. On the one hand, a considerable body of knowledge has been acquired about the role of hormones as "internal" factors controlling growth and development; on the other hand, the profound importance of environmental factors, such as day length and temperature, in the regulation of some of the major phases in the plant life cycle has been clearly demonstrated, although there is considerable evidence that a number of environmental influences are mediated through effects on the levels and distribution of hormones within the plant.

It is axiomatic that the plant body at any given stage is the resultant of the interaction between the inherent (genetic) potentialities of the species and the external factors of the environment. Thus, we cannot say that certain characteristics of the plant are determined genetically, whereas others are environmentally determined, since *all* its characteristics are affected by both genetical and environmental influences. However, it is quite legitimate to say that some *differences* between plants are primarily genetically determined whereas others are due to environmental factors. Thus, the lack of chlorophyll in a plant may be caused by a mutation affecting chlorophyll biosynthesis. On the other hand, a plant may lack chlorophyll because it has been grown in the dark, so that it is etiolated. But it must be emphasized again that the development of a normal green leaf requires both the appropriate genetical factors and certain environmental conditions, including light.

When we speak of the genetical potentialities of the species we must include not only genes located in the nucleus, but also cytoplasmic factors. Certain characters of the plant, including some chloroplast characters, show cytoplasmic inheritance. This fact should not surprise us unduly, since it is now well established that chloroplasts contain DNA and are probably self-replicating organelles. In this book we are not primarily concerned with genetical aspects of morphogenesis, but in all our discussions of this problem it is a basic assumption that, in the final analysis, development involves the expression of the information stored in the genes.

THE LOCALIZATION OF GROWTH

One of the essential characteristics of organisms is that they are able to take up relatively simple substances from their environment and use them in the synthesis of the varied and complex substances of which cells are composed. It is this increase in the amount of living material which is basically what we mean by growth. At the cellular level the increase in living material normally leads to an increase in cell size and ultimately to cell division. These two aspects of growth are seen in their simplest form in unicellular organisms such as bacteria, unicellular algae and protozoa, where growth leads to enlargement of each cell, which then divides and the process is repeated.

When we come to consider the growth of multicellular organisms, such as the higher plants, the situation is much more complex. It is true that here, also, growth ultimately depends on the enlargement and division of individual cells, but not all cells of the plant body contribute to the growth of the organisms as a whole, for growth is restricted to certain embryonic regions, the *meristems*. This restriction of the growing regions is probably related to the fact that mature plant cells are normally surrounded by relatively thick and rigid cell walls, and many cells of mechanical and vascular tissues are, of course, non-living. These facts would probably render co-ordinated growth, involving both cell division and cell enlargement, difficult in an organ such as a stem, once a certain stage of differentiation had been reached. We shall see later that most living plant cells retain the capacity to divide under certain conditions, but even if they do divide the daughter cells do not necessarily increase in size, unless they are relatively thin-walled cells which are able to revert to the embryonic or "meristematic" condition. In having rather strictly localized embryonic regions higher plants differ from animals, where growth typically occurs throughout the organism as a whole.

This difference between higher plants and animals is no doubt related to the basic differences in the modes of nutrition of the two groups. Because they have to take up water and mineral salts from the soil, the autotrophic land plants must necessarily be rooted and sessile, whereas most animals have to forage for their food, whether they are herbivorous or carnivorous, and they need, therefore, to be mobile. This requirement for mobility in animals which forage for their food, in turn, demands that they should have flexible bodies, whereas the plant body can be much more rigid and indeed it needs to be so in erect-growing plants, especially in large forest trees. This rigidity and firmness of the plant body depends upon the presence of relatively thick and firm cell walls, whether in the living cells of the leaf, for example, or in the non-living cells of mechanical tissue of the stem. (The rigidity of those tissues consisting mainly of living cells depends, of course, on the turgidity of the cells and not simply on the mechanical properties of the walls, but even in such tissues a cell wall is an essential requirement for the attainment of the turgid condition.) On the other hand, in aquatic plants, whether they are lower plants or angiosperms, nutrients may be absorbed from the surrounding water directly into the shoot, so that they may be free-floating, and the mechanical tissues are usually less well developed than in land plants.

A number of different types of meristem may be recognized in the plant body. The axial organs, the stems and roots, have *apical meristems*, i.e. growth in length is restricted to the tip regions and the new tissue is added to the plant body on the proximal side, so that the pattern of growth may be described as *accretionary*. The apical meristems of the stem and root usually remain permanently embryonic and capable of growth over long periods—for hundreds of years in some trees. Consequently we may describe these as *indeterminate* meristems.

On the other hand, other parts of the plant, particularly the leaves, flowers and fruits, show rather different patterns of growth and they are embryonic for only a limited period before the whole organ attains maturity. Thus, the growing regions of such organs are sometimes referred to as *determinate* meristems. In such organs the pattern of growth resembles that of animals in that, firstly, there is an embryonic phase of limited duration and secondly, in such organs growth is more generalized than in stems and roots.

The presence of indeterminate meristems, together with the capacity for forming branches, each with its apical meristem, gives the plant body a much less precise and definite form than is the case for the animal body. Indeed, the general form of the plant body resembles a colony of coelenterates, such as corals, rather than that of an individual higher animal. On the other hand, the organs showing determinate growth, such as leaves and flowers, generally show much more precise morphology and may have fairly precise numbers of parts, such as petals.

In addition to classifying meristems as indeterminate and determinate we may classify them in various other ways. For example, we may distinguish the apical meristems of stems and roots, from the *lateral meristems*, comprising the cambium and phellogen (cork cambium). In some plants there are *intercalary meristems*, inserted between regions of differentiated tissues. One of the best known examples of this type of meristem is seen in grasses, where the internodes and leaf sheaths continue growth in the basal region, after the upper parts have become differentiated. The structure of some of these meristems will be described in more detail in Chapter 2.

CELL DIVISION AND CELL VACUOLATION

The growth of a multicellular plant involves both increase in cell number, by cell division, and increase in cell size. These two aspects of growth have no sharp spatial boundaries; however, in the apical regions of shoots and roots cell division occurs most intensively towards the extreme tip of both organs, whereas the region of most rapid increase in cell size is in a zone a few millimetres back from the tip (Fig. 1). In organs of determinate growth, such as leaves and fruits, these two aspects of growth tend to be separated in time, so that there is an early phase in which cell division is predominant, followed later by a phase when cell division ceases and there is active increase in cell size. The greater part of this increase in size is due to vacuolation, i.e. by water uptake, and as a result the cytoplasm may come to be limited to a thin boundary layer against the cell wall.

In the tip regions of roots and shoots in which cell division predominates, the cells are relatively small, and have prominent, spherical nuclei lying towards the centre of the cytoplasm, which is non-vacuolated and tends to be densely staining; the cell walls are thin. The details of the process of mitosis by which the nucleus divides, need not be described here. As a result of division, each of the two daughter cells is only half the size of the parent cell. These cells then proceed to enlarge, but such cell growth involves the synthesis of cytoplasm and cell wall material and not vacuolation.

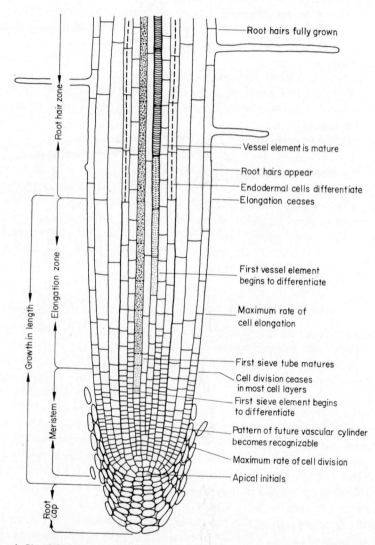

FIG. 1. Simplified diagram of the growing zone of a root, in longitudinal section. The number of cells in a living root is normally much greater than is shown in this diagram. (Reprinted from P. M. Ray, *The Living Plant*, Holt, Rinehart & Winston, New York and London, 1963.)

Since the number of cells in the zone of cell division tends to remain fairly constant (at least over limited periods), it is clear that not all the daughter cells formed in this zone retain the capacity for unlimited further division. The situation is perhaps best illustrated by reference to plants which grow by a single apical cell, such as certain algae and the bryophytes and some pteridophytes where it can clearly be seen that division of the apical cell results in one cell on the outside which becomes the new apical cell, and a second daughter cell, on the proximal side, which gives rise to the differentiated tissue of the thallus or shoot (Fig. 34). This latter daughter cell usually undergoes several further divisions but ultimately the derivative cells lose their capacity for division. Thus, whereas the apical cell remains permanently meristematic, the derivative cells are capable of only a limited number of further divisions. The situation must be analogous in the more complex apices of gymnosperms and angiosperms, where there is normally a number of initial cells i.e. cells which remain meristematic and undergo repeated division, but it is more difficult to recognize which of the daughter cells is destined to remain meristematic and which will give rise to mature tissue. The problem of why cells in the initial zone remain permanently embryonic or meristematic, whereas the derivative cells on the proximal side are capable of only a limited number of further divisions is an intriguing one, but it remains unsolved at the present time.

At a certain distance from the apex, in both shoots and roots, the process of vacuolation commences and, as a result of this process, the root cells of *Allium* may increase in length from 17 μ to 30 μ and in volume by 30-fold. In other tissues the cells may increase up to 150-fold in total volume during vacuolation. It appears that this great uptake of water during cell extension is essentially governed by osmosis, and if we apply the usual concept relating to water uptake by cells, then, in general, the ability of the cell to take up water is given by the diffusion pressure deficit (DPD) or "suction pressure" (SP) which is equal to the osmotic pressure (O) of the vacuolar solution minus the wall or turgor pressure (T). That is, DPD = O − T. Now clearly water uptake may involve either an increased osmotic pressure, or a decreased wall pressure, or both. Studies on the changes in osmotic pressure of the vacuolar solution during growth have yielded no evidence of an increase in osmotic pressure. Indeed, since the vacuolar sap becomes greatly diluted during growth, considerable amounts of additional osmotically active substances, such as sugars, salts, organic acids, etc., must pass into the vacuole during growth, in order simply to maintain the osmotic pressure at a steady value. In some organs the osmotic pressure of the vacuole may actually fall during this phase of growth. Thus, in the petioles of the water lily, *Victoria regia*, which may increase in length from 9 cm to 68 cm in 24 hours, the osmotic value may fall to less than half its original value during the extension phase. On the other hand, there is considerable evidence that in vacuolating cells the wall pressure is reduced by increased

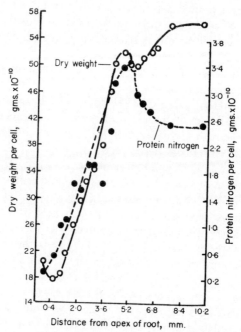

FIG. 2. Changes in dry weight and protein content of cells at increasing distances from the apex of pea roots. (Adapted from R. Brown and D. Broadbent, *J. Exp. Bot.* **1,** 249–63, 1950.)

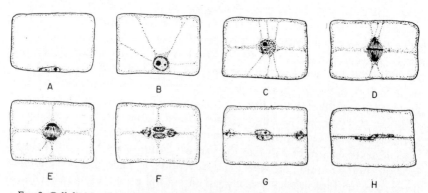

FIG. 3. Cell division in vacuolated cells. A, interphase; B, early prophase; C, prophase; D, metaphase; E, anaphase; F and G, telophases; H, two daughter cells at interphase. (From E. W. Sinnott and R. Bloch, *Amer. J. Bot.* **28,** 1941.)

plasticity of the cell wall at this time (p. 91). As a result of its increased plastic extensibility the wall undergoes irreversible elongation during vacuolation.

Although the greater part of the increase in cell volume during vacuolation is due to water uptake, the synthesis of new cytoplasm and cell wall material proceeds actively during this period, so that the cell increases considerably in dry weight (Fig. 2). Thus, the processes of cell growth initiated before vacuolation commences are continued during this latter phase. Moreover, the zones of cell division and cell vacuolation are not sharply demarcated, and in both shoots and roots of many species cell division occurs in cells which have started to undergo considerable vacuolation (Fig. 3). Division may also occur in vacuolated cells in wound tissues. In root tips the separation of the zones of division and vacuolation are somewhat sharper and division in vacuolated cells is less frequent.

Since growth involves various endergonic, i.e. energy-requiring processes, including protein synthesis, it is not surprising to find that rapidly elongating tissues of the root have a high respiration rate, when compared with mature tissues on the basis of equal *volumes* of tissue, although when expressed *per cell* the respiration rate of mature cells may be greater than that of meristematic cells, since the latter are smaller and contain less cytoplasm. Moreover, growth requires aerobic conditions and an adequate supply of carbohydrate, both as an energy source and as structural material.

The role of growth hormones in cell division and cell extension will be discussed later.

GROWTH OF CELL WALLS

During cell extension the area of the cell wall may increase greatly and this fact poses a number of problems. It might be expected that as the wall is stretched by turgor pressure, it would decrease in thickness, but usually this does not occur. Hence, new material must be added to the wall during growth. There has long been a dispute as to whether the new material is added by "intussusception" throughout the thickness of the wall, or whether it is added to the interior surface, i.e. by "apposition". The bulk of the evidence now supports the second view, at least for many types of cell, but the possibility that there is also some intussusception cannot be excluded. Before we can consider the problem of wall growth further, however, it is necessary to consider the structure of the wall.

Electron microscope studies have shown that the main structural element of the wall in higher plants consists of a framework of cellulose *microfibrils* (Fig. 4.1), which are somewhat flattened in cross-section, having a width of 10–30 nm,* a thickness of 5–10 nm and a length of at least 60 nm. The cellulose of the microfibrils is mainly present in a crystalline state, i.e. the molecules are regularly

* nanometers $= 1 \text{ m} \times 10^{-9}$.

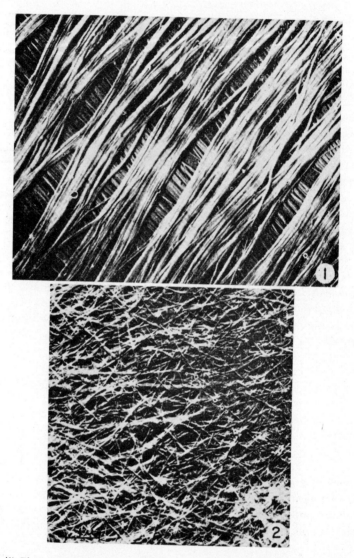

FIG. 4. (1). Electron micrograph, showing structure of the secondary wall of *Valonia* ×
7000. (2). As above, but of the primary wall × 8000. (From F. C. Steward and K.
Mühlethaler, *Ann. Bot.* N.S., **17**, 295, 1953.)

arranged in a lattice, while the remainder is semi- or para-crystalline. The microfibrils are embedded in a continuous matrix, consisting mainly of the so-called *hemicelluloses* (non-cellulosic polysaccharides, composed mainly of residues of the pentoses, arabinose and xylose, and the hexoses, glucose, galactose and mannose) and *pectins*, which contain a high proportion of galacturonic acid residues. The matrix also contains low amounts of proteins and lipids. (Further details of the composition of cell walls are given in Chapter 5, pp. 93, 94.)

Growth of the wall involves the yielding of the wall to the stress generated by turgor pressure. During the extension of the walls the microfibrils become reoriented. In a typical parenchymatous cell undergoing elongation, the microfibrils are at first oriented in a transverse direction (i.e. at right angles to the long axis of the cell), but as the wall becomes stretched they may be arranged

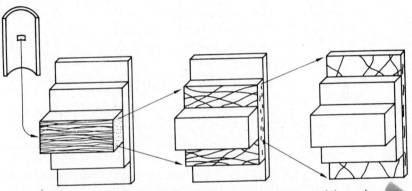

Fig. 5. The multinet concept of cell wall growth showing (left to right) re-orientation of microfibrils as successive stages of wall extension. (From P. A. Roelefsen, *Adv. in Botanical Research*, **2**, 69–150.)

predominantly along the longitudinal axis. During growth, however, new transverse microfibrils are added to the inside of the wall, so that in a cross-section of the wall we find a gradual transition from transversely to longitudinally oriented microfibrils, in passing from the inside to the outside (Fig. 5).

The increased plasticity of the cell wall during vacuolation, referred to earlier, must indicate that the various types of chemical bond which link the different wall components must be temporarily broken during wall growth, as the result of the activities of hydrolytic enzymes (p. 95).

In many types of cells, growth occurs fairly uniformly over the whole wall, giving the so-called "multi-net" pattern of growth. In other cases, as in root hairs and pollen tubes, the cell may extend by "tip growth"; in such cases it is found that in the growing tip region of a cell the microfibrils are oriented in a random fashion (Fig. 4.2), but during the process of wall stretching they become predominantly oriented in the direction of the cell axis. (Figure 4.2

relates to the primary wall of the alga, *Valonia*, but a similar structure is found in the tip region of cells which show tip growth.)

It is not known what determines the initial transverse arrangement of the microfibrils, but it is found that they usually lie parallel to certain *microtubules* which, as the name suggests, are elongated cylindrical structures of diameter 23–27 nm, found in the boundary layers of the cytoplasm. (Fig. 6). Moreover, treatment with colchicine, which disrupts the microtubules, also disorganizes the arrangement of the microfibrils, but does not prevent their deposition. Thus, the orientation of the microtubules may, in some way not yet understood, determine that of the microfibrils of the wall.

The Golgi bodies also appear to play a role in cell wall synthesis, since they are conspicuous in regions of active wall synthesis, especially during the development of the cell plate following division (see below). Moreover, vesicles formed by the Golgi bodies have been shown to contain polysaccharide material. It is possible that the Golgi bodies are responsible for the deposition of the hemicelluloses and pectins of the matrix of the cell wall.

THE FORMATION OF NEW CELL WALLS

Cell division involves the formation of a new cell wall between the two daughter cells. The process commences with the appearance of large numbers of vesicles in the plane of the equator of the spindle. These vesicles are apparently formed by Golgi bodies, and may contain polysaccharides from which the first stage of the new wall, known as cell plate, is formed by coalescence of the vesicles. The cell plate forms first in the centre of the cell and its edge extends outwards, apparently by the addition of vesicles from Golgi bodies which lie on the periphery of the plate, until it joins up with the lateral walls (Fig. 3).

Since the new wall is formed at the equator of the spindle, the plane of the wall is determined by the orientation of the spindle. Recent electron microscopic studies on dividing cells in the roots and coleoptiles of wheat suggest that the orientation of the future spindle is determined by certain changes occurring in the cytoplasm before the nuclei enter prophase. In resting cells the microtubules lie in the outer cytoplasm, just inside the plasmalemma. In cells which are about to undergo division, but in which the nucleus has not yet entered prophase, the "wall microtubules" just described disappear and a band consisting of a large number of microtubules appears in the outer cytoplasm near the longitudinal walls, and at right angles to the axis of the cell. The band appears to run right round the wall surface, in the mid-region of the cell (Fig. 6).

In cells which divide equally, the plane and position of the preprophase band of microtubules coincide with those of the future new wall, but certain of the epidermal cells of the roots of *Phleum pratense* divide unequally by the asymmetric positioning of the new cross wall (p. 54), and in such cells the band of

microtubules still appears in the mid-region of the cell. It would appear, therefore, that the band is not primarily concerned with the location of the cell plate,
but with the orientation of the spindle, though how this is achieved is not
known.

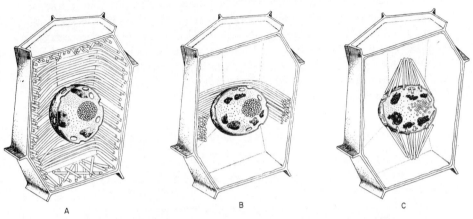

A   B   C

Fig. 6. Diagram of changes in microtubules at cell division. (Prints supplied by Dr.
Myron Ledbetter, Reproduced from *Symposium Internat. Soc. Cell. Biol.*, **6,** 1967.)

MEASUREMENT OF GROWTH

So far we have been concerned largely with qualitative and descriptive
aspects of plant growth. It is also important to study growth quantitatively,
however, and for this purpose we need methods for measuring growth.

As we have already seen, in the final analysis, we might say that growth
involves an increase in the amount of living material. However, it is not always
easy to measure this increase in living material without destroying the organism
in the process. Moreover, if we simply include the protoplast material (cytoplasm and nucleus), we shall leave out increase in such materials as cell walls
which form an integral part of the plant body.

It is possible to adopt a different approach to the problem. Since growth
essentially involves increase in cell number, we may use this criterion as a
measure of growth. Thus, we can measure the growth of a colony of unicellular
organisms by counting the increase in the number of individual cells.

In multicellular organisms, such as higher plants, growth still involves large
increases in cell number, but it clearly is inconvenient, if not impossible, to
measure such increases. However, this increase in cell number, which is accompanied by cell growth, leads to an increase in *size* and in the case of a root or an
unbranched shoot it may be convenient to measure simply the increase in

length or height over a given interval of time. This method is not usually appropriate for a complex root or shoot system, however. Since, if we are studying the growth of a whole plant, we are concerned with the increase in total new tissue formed, it is frequently most appropriate to study changes in the *dry weight* of the plant, which will reflect the actual amount of new organic material synthesized by the plant. However, even a change in dry weight is not always a satisfactory measure of growth, since plant tissues may increase in dry weight due to the accumulation of reserve materials, such as starch and fat, although they may not be growing. Conversely, a germinating seed may show an overall loss in dry weight, due to the utilization of reserves in respiration, although there is no doubt that it is growing.

COLONY GROWTH IN MICRO-ORGANISMS

Before considering the growth curves of higher plants, it is useful to study the growth of a colony of unicellular organisms, such as bacteria or yeast, which multiply by division or "budding", or of a multicellular organism, such as duckweed (*Lemna*), which similarly multiplies by a form of budding.

Consider the growth of a colony of bacteria maintained under constant nutritional and environmental conditions, so that there is a constant rate of cell division. Assume also that the cells divide synchronously, i.e. all cells in the colony divide simultaneously. (Synchronous division can be achieved with cultures of certain organisms.) If the initial number of cells in the colony is n_0, and n the number of cells after a given number of divisions, then,

$$\text{at the end of the 1st generation } n = n_0 \times 2$$
$$\text{,, ,, 2nd ,, } n = n_0 \times 2 \times 2$$
$$\text{,, ,, } x\text{th ,, } n = n_0 \times 2^x$$

This latter relation indicates that the number of cells in the colony is increasing by geometric progression or "exponentially", i.e. at an ever-increasing rate, and if we plot n against the number of generations, we obtain a curve of the form seen in Fig. 7A.

We can rewrite the equation $n = n_0 \times 2^x$ as:

$$\log n = \log n_0 + x \log 2 \qquad (1)$$

It will be seen that we have an equation expressing the relation between the number of cells in the colony n, and x, the number of generations which have occurred, but normally we require the relation between n and t, the time. Now, if t is the time taken for x generations, and the time of one generation (i.e. time between two successive divisions) is g, then $x = t/g$.

Substituting in equation (1), we get

$$\log n = \log n_0 + t/g \log 2$$

Now   log $2/g$ is a constant (k)

Therefore, $\log n = \log n_0 + kt$ (2)

Now equation (2) is a linear equation of the form $y = a + bx$, where $\log n_0$ corresponds to a, and k corresponds to b. Hence if we plot the *log* of the number of cells present in the colony after different times, against t, we should obtain a straight line. This relation is, in fact, found to hold in practice for various organisms growing under constant conditions, whether they multiply by fission, as in bacteria, or by budding, as in yeast and duckweed (*Lemna*) (Fig. 7B). A colony growing in this manner is said to be increasing "logarithmically" or "exponentially".

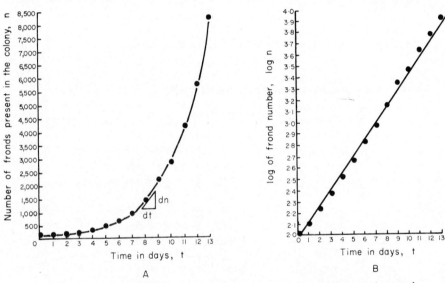

FIG. 7A. Growth curve for a colony of duckweed (*Lemna*) budding synchronously at a constant rate. The initial number of cells is assumed to be 10. B. Growth of a colony of *Lemna* in culture. Linear relationship between logarithm of frond number (log n) and time. (Data from E. Ashby and T. A. Oxley, *Ann. Bot.* **49**, 309, 1935.)

If we consider the type of curve shown in Fig. 7A, giving the increase in the number of *Lemna* "fronds" in the colony, then the *growth rate* of the colony at any time is given by the increase (dn) in the number of cells over a short interval of time, dt, or we can say, growth rate = dn/dt.

The value dn/dt represents the *slope* of the curve at any given time, t, and it will be seen from the graph that the value of the slope increases progressively with time. If all cells are dividing at the same rate, r, clearly at any time, t, the rate of growth of the colony is proportional to the number of cells present, i.e. $dn/dt \propto n$. Thus, although the rate of cell division (r) remains constant, the *absolute growth-rate* of the colony as a whole does not, since as time goes on the number of cells present in the colony increases. The value of r is clearly given by dn/dt (the increase in cell number over a given short interval of time), divided by the number of cells so dividing, i.e.

$$r = \frac{dn}{dt} \cdot \frac{1}{n}$$

and this value is known as the *relative growth rate* of the colony. Thus, for a colony showing this type of growth, the absolute growth rate increases with time, but the relative growth rate remains constant.

It has been shown above (equation (2)) that

$$\log n = \log n_0 + kt$$

This equation can also be written in the form

$$n = n_0 e^{kt} \tag{3}$$

where e = the base of natural logarithms (2·7182).

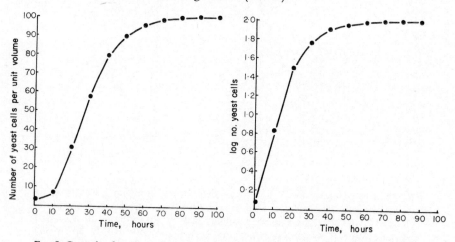

FIG. 8. Growth of a colony of yeast growing in a constant volume of culture solution *Left:* Sigmoid growth curve obtained when number of cells (n) is plotted against time. *Right:* Plot of logarithm of cells (log n) against time. (Data from O. W. Richards, *Ann. Bot.* **42**, 271, 1928.)

In practice, unrestricted growth of a colony can never proceed indefinitely and some limiting factor, such as deficiency of nutrients, must always lead to a decline in growth rate sooner or later. Under cultural conditions in a flask or tube, for example, the food supply will ultimately be exhausted and growth will finally cease. Instead of the typical "exponential" growth curve for cell number, we obtain a "sigmoid" type of curve (Fig. 8 *left*), in which the growth rate increases up to the point of inflection and then declines gradually to zero. When log *n* is plotted against time, growth follows a straight line initially, but later declines (Fig. 8 *right*). In addition to the exhaustion of some food factor, growth in colonies may also be limited by some toxic substance which is formed during growth. The production of such "staling factors" often occurs in cultures of bacteria, fungi, *Chlorella*, etc.

GROWTH OF MULTICELLULAR ORGANISMS

1. *The Exponential Phase*

The sigmoid type of growth curve observed for colonies of unicellular organisms is characteristic also of the growth of individual multicellular plants. This is true not only for the whole plant (Fig. 10), but also for individual organs,

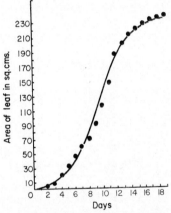

FIG. 9. Sigmoid growth curve of leaf of cucumber (*Cucumis sativa*). (From F. G. Gregory, *Ann. Bot.* **35,** 93, 1921.)

such as leaves (Fig. 9) or internodes. Initially the organism is increasing in size (or weight) by geometrical progression or exponentially. V. H. Blackman (1919) showed that during this initial phase the growth of seedlings follows a "Compound Interest Law" fairly closely and is given by the equation

$$W = W_0 e^{rt} \tag{4}$$

where W = weight of plant after time t,
$\quad W_0$ = initial weight of plant,
$\quad r$ = percentage (or proportional) rate of increase,†
$\quad e$ = exponential coefficient (= 2·7182).

† The same symbol, *r*, has been used here as for the rate of cell division in a colony of bacteria, described above, since in both cases *r* represents the relative growth rate, whether measured by rate of cell division or increase in dry weight.

This equation is clearly exactly comparable with equation (3) above for colony growth and may be derived in a precisely similar manner. From equation (4) we may write

$$\log W = \log W_0 + rt \log e$$

This equation is again of the form $y = a + bx$. This means that we should obtain a straight line when we plot log weight against t (at least for the initial phase of growth, which we are now considering) and this has, in fact, been demonstrated in a number of cases.

From equation (4), it is clear that the final weight attained will depend upon (1) the initial weight, (2) the rate of "interest" (r), (3) the time. The rate of

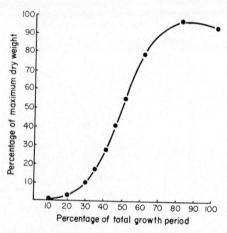

Fig. 10. Increase in dry weight of barley plants during growing season. The points on the curves were derived from smoothed curves drawn from the transformed original data. (From F. G. Gregory, *Ann. Bot.* **40**, 1, 1926.)

"interest" represents the efficiency of the plant as a producer of new material and was called by Blackman the *efficiency index* of dry weight production. A small difference in the efficiency index between two plants will soon make a marked difference in the total yield, and the difference will increase with the lengthening of the period of growth.

It should be noted that the efficiency index is merely a different method of expressing the *relative growth rate* ($dW/W.dt$) as described for colony growth. Whereas the efficiency index (or relative-growth rate) remains constant through the exponential growth phase, the *absolute* increments per unit time increase progressively. The absolute growth increment over a time interval dt is clearly

$$\left(W \times \frac{r}{100}\right).dt.$$

Thus, the absolute growth rate at any given time is proportional to the size of the plant at that time.

The physiological basis of this latter conclusion is easily understood, for when photosynthesis has become active in a young seedling, the power of the plant to synthesize new material (and hence increase in dry weight), is clearly dependent upon its leaf area. Therefore, as the plant grows and increases its leaf-area, the rate at which new material is assimilated will increase proportionately.

2. *Later Phases of Growth*

Just as the growth rate (dn/dt) of a bacterial colony ultimately falls off with time due to the exhaustion of nutrients or the accumulation of toxic products, so the growth rate of a multicellular organism decreases gradually, resulting in a

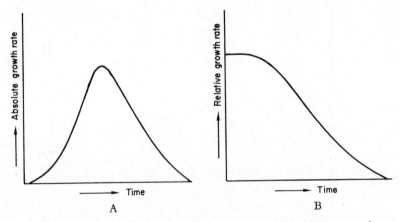

Fig. 11. Changes in (A) absolute growth rate dW/dt and (B) relative growth rate ($dW/W \cdot dt$), where the dry weight changes follow the type of curve shown in Fig. 10.

sigmoid growth curve. The absolute growth rate (dW/dt) is clearly given by the slope of the growth curve at any time t and if we plot the changes in this growth rate with time we get a curve of the type shown in Fig. 11A. We see that the growth rate attains a maximum (corresponding to the point of inflection of the S-shaped growth curve) and then falls away to zero. If, on the other hand, we plot the relative growth rate ($dW/W \cdot dt$) against time, we frequently get a curve of the type in Fig. 11B. It is seen that the relative growth rate (RGR) remains nearly constant at first but later begins to decline. The changes in RGR with time vary a great deal from one species to another and with the conditions under which the plants are growing. Sometimes it is found that the RGR declines steadily from the commencement of growth, so that a true exponential phase does not occur.

The reason for the fall in the RGR is not fully understood and various hypotheses have been suggested. The deficiency of some nutritive factor is clearly not the cause, as it is in colonies of unicellular organisms under artificial conditions. It has been suggested, however, that the reason for the departure from "exponential" growth in a normal plant is that part of the plant material formed during growth gives rise to mechanical, vascular and other tissues which do not directly contribute to further synthesis of new material. The leaves which are the organs most directly concerned in the synthesis of new material thus constitute a diminishing fraction of the total plant weight, i.e. the ratio

$$\frac{\text{Total leaf area } (L)}{\text{Total dry weight } (W)}$$

(known as the leaf/weight ratio, LWR). gradually falls.

Now the relative growth rate (RGR)

$$= \frac{dW}{dt} \cdot \frac{1}{W}$$

Multiplying numerator and denominator by L, we get

$$\text{RGR} = \frac{dW}{L} \cdot \frac{L}{W} \cdot \frac{1}{dt}$$

Now $(dW/L \cdot dt)$ is the *net assimilation rate* (NAR), which is a measure of the rate of increase in dry weight per unit leaf area due to photosynthesis, minus the losses due to respiration; that is, the NAR is a measure of the net efficiency of the plant in production of dry matter. It will be seen that we have the following simple relationship between the three parameters:

$$\text{RGR} = \text{NAR} \times \text{LWR}$$

If we assume that NAR does not change appreciably with the age of the plant, then the fall in RGR must primarily be due to the fall in LWR, for the reasons already indicated. However, both NAR and LW are frequently found to decline during the growth period and thus to contribute to the fall in RGR.

The foregoing simple mathematical relationships have been used to analyse the growth of crops. Thus, the determination of RGRs for different crop plants gives us a useful basis for comparing their growth rates. Similarly, by determining NAR and LWR we can obtain some indication of how differences in RGR arise.

FURTHER READING

General

CLOWES, F. A. L. and B. E. JUNIPER, *Plant Cells*, Blackwell Scientific Publications, Oxford, 1968.

SINNOTT, E. W. *Plant Morphogenesis*, McGraw Hill, New York, 1960.
STEWARD, F. C. *Growth and Organization in Plants*, Addison-Wesley, Reading, Mass., 1968.

More Advanced Reading

BLACKMAN, G. E. Responses to environmental factors by plants in the vegetative phase. Chapter 25 in *Growth in Living Systems* (Ed. M. X. Zarrow), Basic Books, Inc., New York, 1961.
BURSTRÖM, H. Physics of cell elongation. *Encycl. of Plant Physiol.* **14**, 285, 1961.
MÜHLETHALER, K. Ultrastructure and formation of plant cell walls. *Ann. Rev. Plant Physiol.* **18**, 1, 1967.
ROELEFSEN, P. A. Ultrastructure of the wall in growing cells and its relation to the direction of growth. *Advances in Botanical Research*, **2**, 69, Academic Press, London and New York, 1965.
WHALEY, W. G. Growth as a general process. *Encycl. of Plant Physiol.* **14**, 71, 1961.

CHAPTER 2

Patterns of Growth and Differentiation

LEVELS OF DIFFERENTIATION

So far, we have been concerned primarily with growth, rather than differentiation. In the present chapter we shall describe the development of the main organs of the plant and the way in which differentiation arises within the plant.

Now, when we consider the manifold forms of differentiation in the plant it is evident that it occurs at various levels. At the highest level, there is differentiation in the plant body as a whole, as seen in the division into root and shoot. Within the shoot we can observe the differentiation into various organs, such as stems, leaves, buds and flowers, and within each of these organs there is differentiation at the cellular and tissue level. These three levels of differentiation also constitute a series of successive stages in *time*—there is first differentiation into root and shoot in the embryo, and this is followed by the formation of organ primordia, as a result of the activities of the apical meristems. These organ primordia do not at first show differentiation at the cell and tissue level, which occurs during the later stages of their development. We may illustrate the progressive steps of differentiation of the plant body in a diagram (Fig. 12), in which it is seen that at various stages there are divergent alternative pathways, leading to successively more specific pathways of differentiation.

The chain of developmental events usually takes place in a very orderly manner, one stage following another in a proper sequence. This orderly sequence of changes suggests that the successive steps are not controlled independently, but that the attainment of one stage exercises some control over the alternative pathways which will be entered at the next step. This idea is far from new, since it was clearly enunciated by Pfeffer in 1903.

Once a particular pathway of development has been entered, the process is usually irreversible; for example, once a flower primordium has been formed, it is not easily converted back to a leafy shoot. Similarly, cortical cells do not normally change directly into vascular elements nor vice versa. Thus, at certain stages organs and tissues become *determined* in their pattern of development.

Nevertheless, some tissues do retain the capacity for "de-differentiation", so that roots may be initiated in stem tissue, and so on. The successive entry into specific developmental pathways which, once entered, cannot easily be departed from, has been referred to by Waddington as the "canalization of development" (p. 280) (Fig. 122).

In addition to the first major step in differentiation (viz. the formation of root and shoot), certain other changes occur during the life cycle of seed plants, which must be regarded as aspects of differentiation, of which the most important is the transition to the reproductive phase, which involves a profound change in the structure of the shoot apex (p. 45). We shall see later that

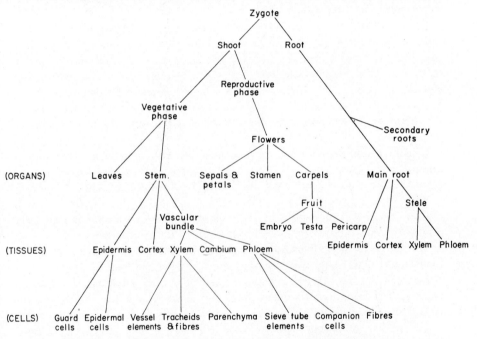

Fig. 12. Levels of differentiation during the development of a higher plant.

in many species the onset of flowering is controlled by environmental factors, but in many other species it appears to be determined more by progressive changes occurring during the development of the plant itself, than by environmental factors. Often these progressive physiological changes are reflected in morphological characters, such as leaf shape, in which a gradient up the stem may frequently be seen. These aspects of development will be dealt with later, but it is important to recognize that they represent an aspect of differentiation within the shoot as a whole.

TISSUE AND CELL DIFFERENTIATION

When the development of the main organs has been initiated, with the determination of stem, leaf, root and flower, each of these organs follows a largely
independent course of further development, as shown by the fact that a leaf
primordium will undergo full differentiation if isolated and maintained in
sterile culture (p. 37). However, in the intact plant development is a continuous
process and it is somewhat artificial to draw too sharp a distinction between
organ and tissue differentiation. In particular, the differentiation of vascular
tissue is continuous throughout the plant and this continuity is no doubt an
important factor in achieving co-ordination of development and function
in the plant as an integrated whole.

At the cellular level, the term differentiation is sometimes used in two
different senses, viz. (1) it may be applied to the development of different
specialized types of mature cell within an organ or tissue; or (2) it may be used
to refer to the changes which occur during the development of a meristematic
cell into a mature cell, usually involving vacuolation and enlargement. In this
book we shall use the term in the first sense and we shall use the term *maturation*
for the processes which lead to the formation of a mature cell from a meristematic one.

Usually maturation involves cell vacuolation and enlargement, and some
aspects of this process have already been described (p. 7). Cells undergoing
maturation may show relatively little other change in structure, as in the formation of parenchymatous tissue, or there may be great changes, as in the formation of xylem and phloem tissue. It is the diverging pathways followed by
different cells during maturation which result in differentiation.

In addition to the visible changes involved in differentiation, there are also
biochemical differences, some of which will be described later (p. 283).

Between the biochemical differences occurring at the molecular level and
the structural changes observable with the optical microscope, one might expect
to find changes at the ultrastructural level, to be seen under the electron microscope. Studies on various types of tissue have shown that, in general, living
differentiated higher plant cells have the normal cell organelles, including
mitochondria, Golgi bodies (dictyosomes), plastids and endoplasmic reticulum,
but there are certain exceptions, e.g. sieve tubes, in which most of the organelles
disintegrate during differentiation. The organelle which shows the most marked
differences in various types of tissue is the plastid, the structure of which varies
enormously according to whether it occurs in leaf tissue, storage tissue, fruits
(as in tomato), or flower parts, such as petals. Mitochondria also vary quite
markedly in number and structure in different types of cell and the Golgi
bodies show active and quiescent states related to the state of wall growth,
secretion and so on. The endoplasmic reticulum varies in abundance and

localization in several types of specialized cell, particularly those concerned with secretion.

On the other hand, many of the differences to be observed between various types of tissue relate to the cell wall. Not only do we find very characteristic differentiation of the cell wall in, for example, the various types of xylem cell, but also in living differentiated cells, such as collenchyma, endodermis and stomatal guard cells. These readily observable differences reflect differences detectable at the ultrastructural level. However, an account of the ultrastructural changes in the plastids and cell wall occurring during differentiation is beyond the scope of this book.

Most cells which have attained the mature condition do not normally undergo subsequent cell division. However, some types of mature cell still retain the capacity to divide, as seen in the origin of vascular cambium, cork cambium and adventitious root initials in various types of parenchymatous tissue. Moreover, many types of cell are able to resume cell division in response to wounding (p. 161).

DIFFERENTIATION INTO ROOT AND SHOOT—EMBRYO DEVELOPMENT

The first major step in differentiation occurs at a very early stage in the development of the embryo, with the establishment of a shoot end and a root end.

As an example of embryo development in angiosperms we may consider that of *Capsella bursa-pastoris* (Fig. 13). The fertilized zygote is a somewhat elongated cell, which divides transversely to give a smaller *terminal* cell, and a larger *basal* cell. The terminal cell gives rise to most of the future embryo, whereas the basal cell gives rise mainly to the *suspensor*. The terminal cell divides by two successive longitudinal divisions, with the plane of the second division at right angles to that of the first, to give four cells; these cells then each divide transversely to form eight cells, which constitute the octant. Each octant cell divides to form an outer protodermal cell, which gives rise to the future epidermis, and an inner cell. The inner cells continue to divide to form the cotyledons and the hypocotyl.

By several successive transverse divisions the basal cell gives rise to a row of cells which forms the suspensor, the end cell of which enlarges and becomes sac-like. The suspensor cell nearest the embryo undergoes several divisions, to give a group of cells, of which the outer ones form the future root cap and root epidermis, while the inner ones form the remainder of the radicle. The fully developed embryo is formed by further repeated divisions of these various regions.

There is considerable variation from the pattern of development described for *Capsella* among the various groups of angiosperms, but the details do not concern us here. However, whatever the variation in further development, the initial stages have certain features in common, namely that the first division of the zygote gives rise to two unequal cells, and of these, the basal cell is normally the one nearer the micropyle of the ovule, and gives rise to the root end of the embryo, whereas the terminal cell gives rise to the shoot end. Thus, even the very young embryo shows *polarity*, in that it has a shoot end

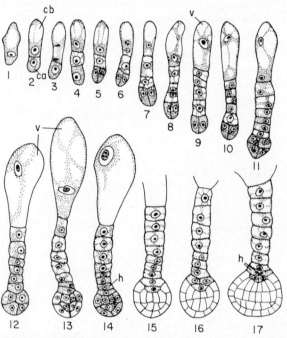

Fig. 13. Early stages in the development of the embryo of *Capsella bursa-pastoris*. Note the initial unequal division into a larger basal cell (cb) and smaller terminal cell (ca). The basal cell gives rise to the suspensor. (From A. Fahn, *Plant Anatomy*, Pergamon Press, Oxford, 1967, adapted from Souèges, 1914.)

and a root end. Indeed, the egg itself shows differences in the density of cytoplasm between its two ends, suggesting that the first unequal division of the zygote is already predetermined by polarization in the unfertilized egg. We shall discuss the basis of this polarity in more detail later.

After the embryo has become differentiated into root and shoot regions, apical meristems are established and some organs become differentiated, often while the seed is still developing on the parent plant, so that not only cotyledons but also a rudimentary epicotyl and, in grasses, even several leaf primordia, may be present.

SHOOT APICAL MERISTEMS

Although this book is primarily concerned with flowering plants, it is useful to consider, briefly, the patterns of growth in the lower plants. In simple filamentous algae, such as *Spirogyra*, every cell appears to be potentially capable of division and growth and is not localized to particular regions. However, in many algae there is marked localization of growth. Thus, the alga, *Chara*, has a single prominent apical cell, which divides repeatedly, giving a larger outer (distal) cell, which continues to function as the apical cell, and a smaller, proximal daughter cell, which proceeds to undergo further division, the resulting cells forming the mature tissue of the thallus (Fig. 34).

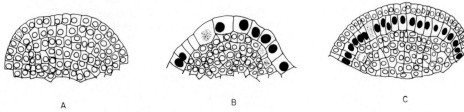

A   B   C

Fig. 14. Longitudinal sections of the shoot apex of *Datura*. A. Normal apex, with all cells diploid. B. Apex after treatment with colchicine, showing polyploid nuclei in outer tunica layer. C. After treatment with colchicine, with polyploid nuclei in inner tunica layer. (Adapted from S. Satina, A. F. Blakeslee and A. G. Avery, *Amer. J. Bot.* **27**, 895, 1940.)

In the bryophytes and many pteridophytes, also, the shoot grows by a single well-marked apical cell, which normally has the form of an inverted tetrahedron, and it divides so that the three "inner" faces cut off daughter cells in succession, and these latter cells undergo further division to form the tissues of the shoot.

The early plant anatomists of the nineteenth century were so impressed with the essential unity of structure in vascular plants that they expected to find single apical cells also in gymnosperms and angiosperms and indeed described such cells. Later, however, it became apparent that there is no clearly recognizable *single* apical cell in the shoots of higher plants, but two zones may be distinguished in the shoot apical region of flowering plants: (1) the outer *tunica* or mantle, which surrounds and envelops, (2) the inner *corpus* (Fig. 14A). These zones can be distinguished fairly sharply by the predominant planes of cell division. In the tunica the divisions are predominantly *anticlinal*, i.e. with the axis of the mitotic spindle parallel to the surface, so that the resulting cross-wall separating the two daughter cells is perpendicular to the surface. The corpus, on the other hand, is characterized by the fact that divisions occur in all planes, viz. both anticlinal and *periclinal* (i.e. the spindle is perpendicular and the new wall parallel, to the surface). The thickness of the tunica is somewhat variable and it may consist of one, two or more layers of cells, according to the species. Moreover, even within a single species the number of tunica layers may vary according to the age of the plant, the nutrient status and various other conditions.

It should be noted that the tunica–corpus theory is largely descriptive and simply based on what can be observed—it makes no predictions regarding the future destiny of the tunica and corpus cells. The epidermis does, of course, arise from the outer layer of the tunica, but there is considerable variation between different species in the extent to which the two layers may contribute to the origin of leaves and buds (see p. 30). However, although we should not regard the tunica and corpus as rigidly and permanently demarcated, it is possible to show that in some species the outer tunica layers remain remarkably distinct from the deeper tissue for long periods. One method by which this has been

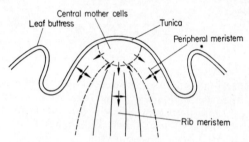

Fig. 15. Generalized diagrams to illustrate zonation in the shoot apical region of a flowering plant.

demonstrated is to induce polyploidy in the cells of the shoot apices of *Datura* and of maize by treatment with colchicine. The tetraploid cells so formed can be recognized by their larger nuclei. If tetraploid cells are formed in the outer layer of the tunica, for example, all the resulting daughter cells will show large nuclei and it can be seen that such cells are strictly limited to a single layer of the tunica (Fig. 14B), indicating that periclinal divisions are very rare and that each layer of the tunica retains its identity remarkably constantly. Each of the layers of the tunica arises from a set of initial cells at the shoot apex, and the corpus apparently arises from its own set of initials, although there is some difference of interpretation on this point.

The distinction between tunica and corpus is even less definite in gymnosperm shoot apices, and although the divisions in the outer layers of *Pinus*, for example, are predominantly anticlinal, there are also quite frequent periclinal divisions as well.

In addition to the tunica and corpus and their initials, it is possible to recognize a number of other zones in the apices of some species. One simple form of zonation is shown in Figs. 15 and 16. At the summit of the apical dome there is a group of initial cells which divide mainly anticlinally, and so give rise to the tunica, but the lower layers may also divide periclinally. Below these initial cells there is a group of larger cells, known as the *central mother cells*, which appear to have a low rate of division. Divisions at the boundary of the central mother zone give rise to (1) a zone of actively dividing cells which form the flanks of the apex and is sometimes called the *flank* or *peripheral meristem*; and (2) a zone of cells which divide mainly along the shoot axis and give rise to the longitudinal rows of cells of the cortex and pith and hence is sometimes called the *rib meristem*. The distinction between these zones, which have been given various names by different workers, is somewhat arbitrary, since their boundaries are ill defined and grade into each other. Moreover, there is considerable

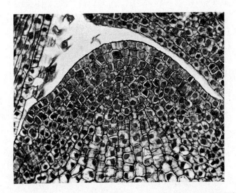

FIG. 16. Longitudinal section through vegetative shoot apex of *Chrysanthemum mori folium*. The various zones shown in Fig. 15 may be distinguished. (From R. A. Popham, *Amer. J. Bot.* **37**, 476, 1950.)

variation between species, both with respect to the general shape of the apical region and to the number of zones which can be recognized within it, and yet the apices of all these species produce similar end products, viz. stem, leaves and buds. It would seem, therefore, that not much morphogenetic significance can be attached to the various zones which can be recognized in the shoot apex.

THE INITIATION OF LEAVES AND BUDS

A leaf originates from the periclinal divisions of a group of cells on the flank of the apex (Fig. 21). As a result of these divisions in a localized area, a small protuberance, known as the *leaf buttress* (Figs. 20 and 21) is formed and gives rise to the future leaf primordium. The number of layers involved in these initial divisions varies considerably in different species. In many grasses, the periclinal divisions commence in the outermost layer of the tunica, and in the layer below. In other monocotyledons and in dicotyledons, periclinal divisions do not take place in the outermost layer, but in the layers below it. Thus, the extent to which the tunica and corpus are involved in the initiation of the primordium varies greatly. In many species the initial divisions involve both the tunica and the corpus, while in others they may occur in only one or other of these layers. Variations may occur even within a single species.

Lateral buds usually appear somewhat later than leaf primordia, in the sequence of developmental changes seen at the shoot apex. Buds arise in the outer layers of the stem tissues, as a result of cell divisions which may be predominantly anticlinal in the outer layers, or both anticlinal and periclinal in the deeper layers. As with leaves, there is considerable variation between species in the extent to which tunica and corpus are involved in these cell divisions giving rise to lateral buds. As a result of these cell divisions the bud emerges as a protuberance and it soon develops an apical structure similar to that of the main shoot apex for that species.

THE SITING OF LEAF PRIMORDIA

The siting of leaf primordia at the shoot apex is a rather precisely regulated process, although there are considerable differences from one species to another

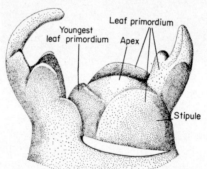

Fig. 17. Vegetative shoot apex of *Vitis*. (From A. Fahn, *Plant Anatomy*, Pergamon Press, Oxford, 1967.)

in the pattern of arrangement of leaf primordia. In considering this problem it has to be remembered that the apex is growing continuously and that as it does so the older leaf primordia are left behind on the flanks of the apex and they steadily increase in size as they do so (Fig. 17). As the apex grows, new leaf primordia are being continuously initiated above the existing ones.

The siting of leaf primordia at the apex determines the arrangement of leaves on the mature shoot, for which the term *phyllotaxis* is used. The most common type of leaf arrangement is *spiral* phyllotaxis, in which it can be seen that a line drawn through successively older leaf primordia at the shoot apex forms a spiral, known as the *genetic* or *developmental spiral* (Fig. 18). The mathematical treatment of spiral phyllotaxis has interested botanists for more than a century, but here we shall consider the problem only so far as it relates to the siting of primordia at the shoot apex.

The fern apex provides very convenient material for studying phyllotaxis, since it is relatively flat and the primordia are well spaced out (Fig. 19). In considering this problem it is useful to use a system of nomenclature in which the youngest primordium is called P_1 and the successively older primordia, P_2, P_3, etc. The next primordium to arise is referred to as I_1 (i.e. Initial$_1$) and the successively *younger* primordia as I_2, I_3, etc. If we draw radii from the centre

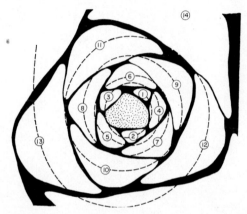

FIG. 18. Diagram to illustrate spiral phyllotaxis as seen in cross-section through shoot apical region of *Saxifraga* (genetic spiral shown by broken line). (From F. Clowes, *Apical Meristems*, Blackwell Scientific Publications, Oxford, 1961.)

of the apex to two successive leaf primordia, it is found that the angle of divergence between the two lines varies from one species to another, but where there are numerous primordia at the apex, it is found to approach a "limiting" value of 137·5°. It is also found that in many species the radial distance of successive primordia increases in geometric progression, indicating that there is a corresponding increase in the rate of expansion of the apex with distance from the centre. Thus, we can reproduce the spiral phyllotaxis seen at a shoot apex by marking points consecutively around a centre at a constant divergence of 137·5° and radially at a distance from the centre which increases in geometric progression. If we then join the successive points, we shall obtain a genetic spiral.

What determines that the next primordium will be formed at a position which will cause an angular divergence from the preceding primordium of approximately 137·5°? There has been a great deal of controversy on this question, but at present there are two main theories to account for the observed facts, which may be referred to as the "Repulsion Theory" and the "Available Space Theory", respectively. Both theories postulate that the positions in which the leaf primordia arise are determined by the positions of older primordia. According to the Available Space Theory (first put forward by Hofmeister in 1865) a certain minimum space between existing primordia and the centre of the apex is necessary before a new primordium can arise. As the apex grows,

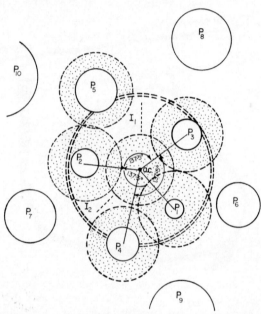

FIG. 19. The growth centre and field concept as it may apply to the apex of *Dryopteris*. The apex is seen from above. *ac*, apical cell; P_1, P_2, P_3, P_4, etc., leaf primordia in order of increasing age; I_1, I_2, the next primordia (as yet invisible) to be formed in that order. The large double-dashed circular line indicates the approximate limit of the apical cone. The subapical region lies outside the double-dashed circle. The hypothetical inhibitory-fields around the several growth centres are indicated by stippling. (Adapted from C. W. Wardlaw, *Growth*, **13**, Suppl. 93, 1949.)

the spaces between existing primordia increase in size and it is postulated that the next primordium (I_1) arises between P_2 and P_3 because this is the first area to reach the necessary minimum size. The next primordium after this (I_2) will arise between P_2 and P_4, and so on.

The Repulsion Theory was put forward by Schoute (1913) who postulated that the centre of a leaf primordium is determined first, and that a specific substance is produced which inhibits the formation of others in the immediate vicinity, so that new primordia again arise in the gaps between older ones, where they will presumably be outside the inhibitory fields of the neighbouring primordia (Fig. 19). Inhibition of new centres by the main apex was also postulated, so that new primordia are prevented from forming within a minimum distance from the summit of the apex. As we shall see later (p. 59), there is, indeed, evidence for the existence of mutual inhibition between growth centres.

These two hypotheses are not necessarily mutually exclusive, since the "available space" may not be determined simply by the superficial area between adjacent primordia, but by freedom from their inhibitory influence. Both theories postulate that phyllotaxis is determined primarily by the geometry of the shoot apex. It can be shown, from purely geometrical considerations, that I_1 will arise at a divergence of 137.5° from P_1, if it occurs at a point which divides the angle between P_2 and P_3 in the inverse ratio of their respective ages; that is, the new primordium will not be equidistant from the neighbouring primordia, but will be displaced towards the older (and larger) of these (Fig. 19). It is not clear what is the significance of this fact, but it is consistent with the hypothesis that the neighbouring primordia play an important part in determining the position of a new primordium.

Attempts have been made to test the "available space" and "repulsion" theories by surgical experiments on the shoot apex. For example, in experiments with *Lupinus albus*, R. and M. Snow made radial cuts in the area at which I_2 would be expected to arise, thereby reducing the space available, and it was found that no primordium developed in this space, presumably because it was reduced below the minimum area necessary for primordium development. On the other hand, Wardlaw isolated I_1 of fern apices by two radial cuts and found that it then grew *more rapidly* than normally (Fig. 22B), suggesting that it had thereby been released from the inhibitory effects of neighbouring primordia. Although these and other surgical experiments have produced interesting results, they have not given decisive evidence in favour of either of the two theories.

DEVELOPMENT OF THE LEAF

The overall development of the leaf may be divided into the following steps: (1) formation of the foliar buttress, (2) formation of the leaf axis, and (3) formation of the lamina. The following account is based upon the development of the tobacco leaf.

As we have seen (p. 29), a small protuberance, known as the *leaf* (or *foliar*) *buttress*, arises on the surface of the flanks of the meristem (Fig. 21) by periclinal divisions in the surface layers. Certain cells towards the centre of the foliar buttress now begin to divide actively and a small finger-like protuberance emerges from the buttress (Fig. 20). This protuberance proceeds to grow in size by the activities of apical cells until it is about 1 mm long. At this stage the leaf primordium consists of little more than the future axis (midrib and petiole) of the leaf. Soon, however, certain cells on its flanks begin to grow, so that it now acquires a more flattened appearance in cross-section. These dividing cells on the flanks of the mid-rib constitute the *marginal meristems* and they give

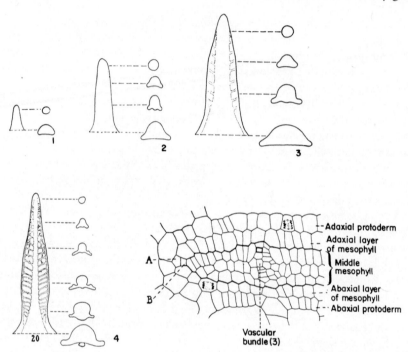

FIG. 20. Diagrams of longitudinal and cross-sections of leaf primordia of *Nicotiana tabacum* at different ontogenetic stages. 1. A young, more or less cone-shaped primordium. 2. Primordium in which the narrow margins, from which the lamina will develop, can be seen. 3. Primordium in which the beginning of development of the main lateral veins can be seen. 4. Primordium 5 mm long in which the early development of the provascular system can be seen. 5. Cross-section of the marginal region of tobacco leaf showing submarginal initial and the origin of the mesophyll and a vascular strand. A, B, submarginal initials. (From G. S. Avery, *Amer. J. Bot.* **20**, 565, 1933.)

rise to the future lamina of the leaf. The tip growth of the mid-rib ceases fairly early, when the total leaf length is 2–3 mm and further growth is more generally dispersed.

The marginal meristems consist of superficial *marginal initial* cells, together with the underlying *submarginal initials* (Figs. 20, 21). In dicotyledons, the marginal initials normally divide only anticlinally, to give rise to the surface layers of the leaf, whereas the submarginal initials form the inner tissues of the future leaf. As a result of these activities of the marginal and submarginal initials a definite number of layers is established in the young lamina and under any given set of environmental conditions this number remains rather constant over most of the leaf, throughout its future development. The relative constancy in the number of layers arises from the fact that the cells continue to divide mainly anti-clinally, i.e. at right angles to the surface of the leaf, so that there is a steady increase in area but not in thickness.

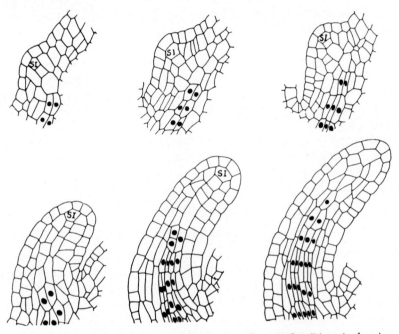

Fig. 21. Early stages in development of leaf primordium in flax (Linum), showing development of procambial strand (cells with nuclei indicated). S.I., submarginal initial cell. (From G. Girolami, *Amer. J. Bot.* **41**, 264, 1954.)

The different layers of the leaf are found to cease cell division at different stages. The cells of the upper epidermis usually cease dividing first, when the tobacco leaf is only 6–7 cm long, i.e. when it is only one-fifth to one-sixth of its final size. However, these epidermal cells continue to increase in size until the whole leaf ceases to enlarge. On the other hand, the palisade cells continue to divide and so give rise to a closely packed layer of cells which keeps pace in its growth with the upper epidermal cells. They cease dividing and enlarging shortly before the epidermal cells do so, so that to keep pace with the latter they are pulled apart slightly in the final stages, thus giving rise to the intercellular spaces between the mature palisade cells. The cells of the future spongy meso-phyll cease dividing and enlarging earlier than the cells of the palisade, so that the cells become pulled apart more in the final stages, giving rise to larger inter-cellular spaces than in the palisade layers and to a more irregular arrangement of the cells. This study of the origin of the different layers of the mesophyll of the leaf provides a good example of how a study of the morphology of development can give a better understanding of the way in which the mature structure of an organ arises.

The first procambium is formed towards the base of the developing leaf primordium at a very early stage (Fig. 21). This first procambium forms the future mid vein and it develops both outwards towards the tip (acropetally) and downward (basipetally) to link up with the procambium of the stem (Fig. 25A). The first vascular elements to appear are those of the protophloem, followed later by the protoxylem. Shortly after, the lateral meristems start to form the lamina, and the first signs of the lateral veins appear when the pri-mordium is about 1·5 mm long. The connecting veins soon appear, towards the tip (Fig. 20).

Complete normal leaf development depends upon exposure to light and is one of the important aspects of "photomorphogenesis" (p. 179). Hormones, especially auxin and cytokinin, also appear to play an important part in leaf growth (p. 106).

LEAF DETERMINATION

The differentiation of the shoot into leaves, buds and stems commences, as we have seen, with the formation of leaf and bud primordia at the shoot apex.

The early stages of development are very similar in both leaves and buds, but whereas the leaf primordium very early assumes a dorsiventrality (i.e. it becomes flattened, with upper and lower surfaces), the bud remains radially symmetrical. Moreover, whereas the apical meristematic cells of the leaf primordium cease to be active at an early stage and the development of the leaf is determinate, the bud primordium develops a typical shoot apical meri-stem which shows indeterminate growth.

Although we know from its position and sequence that a given primordium is normally destined to become a bud or a leaf, a very young primordium is not yet irreversibly determined to become one or the other. Indeed, in the early stages of their development, the primordia which, by their position, would normally give rise to leaves, may be converted into buds by certain surgical treatments. These techniques were first developed by R. and M. Snow and have since been used extensively by others, especially by Wardlaw and his associates, using the fern apex. The interconvertibility of leaf and bud primordia has been demonstrated in *Dryopteris dilatata*, by making a deep tangential cut between the shoot apical cell and a very young presumptive leaf primordium, as a result of which it develops into a radially symmetrical bud instead of a dorsiventral leaf (Fig. 22A). This conversion of a leaf into a bud can only be effected with very young primordia. When a leaf primordium is a little older it remains as a leaf when such surgical treatment is applied. Thus, the primordia produced on the shoot apex are at first uncommitted, and they only become determined as leaves at a later stage.

A different approach to the problem of leaf determination has been made by Steeves using sterile culture methods. When primordia are removed from the apex of the fern *Osmunda cinnamomea* and placed on a sterile nutrient medium they are able to undergo further growth and development (Fig. 23). When the youngest primordia, P_1–P_5, were so tested they developed not as leaves but as shoots which eventually became rooted plants. Progressively older

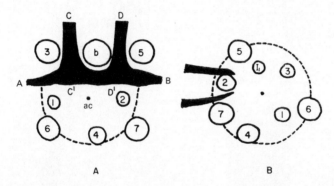

A   B

Fig. 22A. Shoot apex of *Dryopteris* in which the I_1 position was isolated from the apical cells (ac) by a tangential incision (AB) and from leaf primordia 3 and 5 by radial incisions (CC^1 and DD^1). A bud has been formed in what was normally a leaf position. B. Shoot apex in which leaf primordium, P_2, has been isolated from P_5 and P_7 by radial incisions: P_2 grows rapidly and is soon larger than older primordia. (From C. W. Wardlaw, (A) *Proc. Linn. Soc.* **162,** 13, 1950–51; (B) *Growth*, **13,** Suppl'. 93, 1949.)

primordia showed an increasing tendency to develop as leaves, and P_{10} always developed as a dorsiventral leaf. From these experiments, Steeves concluded that leaf primordia of *Osmunda* are not irreversibly determined from their inception, but undergo a relatively long period of development during which they remain undetermined. Determination is gradually imposed on the primordium.

Once determination has occurred, the future development of the complex pattern of the leaf is self-controlled, as shown by the fact that isolated leaf primordia in culture appear to go through all the normal stages of development, even though the resulting leaves are minute.

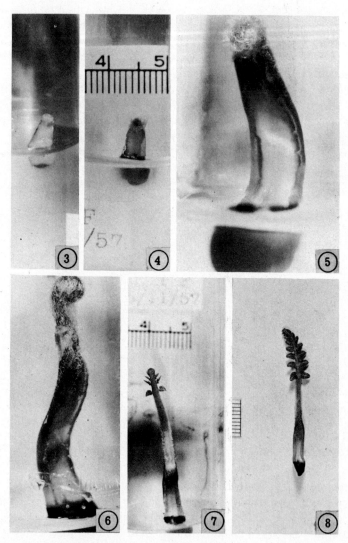

FIG. 23. Sterile culture of isolated leaf primordia of the fern, *Osmunda cinnamomea*, on a simple medium. Successive stages in the development of the leaf, from the earliest stage (top left). (From J. D. Caponetti and T. A. Steeves, *Can. J. Bot.* **41**, 545, 1963.)

LEAF SHAPE

The very wide range of variation in leaf shape in seed plants needs no emphasis. The shape of the mature leaf is determined by three factors: (i) the shape of its primordium; (ii) the number, distribution and orientation of cell divisions; (iii) the amount and distribution of cell enlargement.

The form of the early leaf primordium varies considerably from species to species. As we have seen, in tobacco the young primordium has a simple, finger-like structure, but in maples (*Acer*), the development of the primordium forming the mid-rib is shortly followed by the appearance of two lateral branches at its base, and these three finger-like structures give rise to the main veins of the leaf. In a compound leaf such as that of ash (*Fraxinus*), a number of lateral lobes are formed from the central primordium, and these give rise to the leaflets of the mature leaf. The subsequent development of each leaflet resembles that of a simple leaf.

The comparative growth rates of the lamina and of the main veins have a profound effect on the ultimate form of the leaf. If lamina growth keeps pace with that of the main veins, then a leaf of simple outline results. On the other hand, if growth is more vigorous near the veins than in the other regions, then a lobed leaf will be formed, the ultimate shape depending also upon the pattern of vein development. In maple, for example, the early localized lamina growth around the main veins is normally followed by a wing-like growth to form a continuous sheet of lamina joining the veins, so that a *palmate* leaf is produced, but in some genetical variants the growth of the lamina is restricted more to the regions adjacent to the veins, so that a more "dissected" type of leaf is formed.

Leaf shape may be profoundly modified by environmental factors; for example, the submerged leaves of some aquatic plants, such as *Sagittaria* and *Ranunculus* spp., have a very different form from that of the aerial leaves. In certain species a considerable number of genes which cause wide variations in leaf shape are known. The successive leaves formed on the stem from the seedling stage onwards very commonly show characteristic changes in shape (p. 215).

DIFFERENTIATION IN THE STEM

As we have seen, the cells in the apical meristem itself are generally small, densely cytoplasmic, have large nuclei, and are non-vacuolated. As we pass downwards from the apex to the regions in which cell vacuolation and differentiation begins to be apparent in the pith and cortex, we notice that there is a zone of cells between these two latter regions which is characterized by smaller, deeply staining cells, as seen in cross-section (Fig. 24). This latter zone gives rise to the future procambium. The level at which the procambium can be recognized varies considerably from species to species, but it commonly can be recognized in the zone of leaf initiation. At the highest level these densely staining smaller cells may be seen to form a complete ring, but not all the cells of the ring are destined to form the future vascular tissues and lower down the stem it can be seen that the ring has become broken into discrete strands, by vacuolation of some of the intervening cells of the former ring. These strands constitute the first clearly delimited procambium. Although the cells of the strands appear relatively small in transverse section, in longitudinal section they can be seen to be elongated and spindle-shaped.

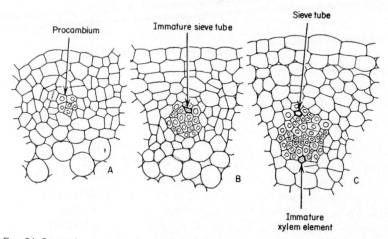

FIG. 24. Successive stages in the development of the procambium (cells with nuclei) in transections of a stem of *Linum perenne*. (All × 430. From K. Esau, *Amer. J. Bot.* **29,** 738, 1942.)

As we have seen (p. 36), the development of the procambium is closely associated with leaf development. From the leaf base the procambium develops both acropetally into the leaf and basipetally to connect up with other vascular strands in the stem (Fig. 25A).

THE SHOOT APEX AS A SELF-DETERMINING REGION

The observation that procambial strands develop acropetally into leaf primordia at a very early stage of their development raises the question as to whether the procambium plays a role in determining the initial siting of leaf primordia. If this were the case, then it would imply that the activities and organization of the shoot apex are influenced and controlled by the already differentiated regions of the shoot. However, several lines of evidence seem to argue against this conclusion.

Firstly, it is clear that an organized shoot apex must arise *de novo* during the development of the embryo, and this may occur in free cell cultures in which

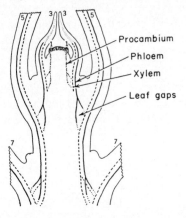

Procambium

Phloem

Xylem

Leaf gaps

FIG. 25A. Diagram illustrating the initial vascular differentiation in a shoot with a decussate leaf arrangement, as seen in longitudinal section. Continuous lines indicate phloem and broken lines indicate xylem. (From K. Esau, *Plant Anatomy*, John Wiley & Sons, Inc., New York, 1953.)

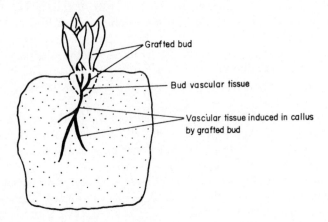

Grafted bud

Bud vascular tissue

Vascular tissue induced in callus by grafted bud

FIG. 25B. The induction of vascular tissue in chicory (*Cichorium intybus*) callus by a chicory shoot bud implanted on the upper surface. The regenerated vascular tissue in the callus becomes connected with that of the bud itself. (After G. Camus, *Compt. Rend. Acad. Sci. Paris*. **219**, 34, 1944.)

the developing embryo is independent of the influence of any surrounding differentiated tissues (p. 156). Similarly, shoot buds may arise spontaneously from undifferentiated tissue; for example, if chicory (*Cichorium*) or dandelion (*Tarax-*

acum) roots are cut in pieces, under appropriate conditions they will regenerate shoot buds from the cut surface at the upper end (Fig. 31.). Adventitious buds will also develop in callus cultures of various species (p. 156). These observations suggest that the shoot apex represents a stable configuration of cells which will, as it were, "crystallize out" from an undifferentiated mass of tissue under appropriate conditions. Thus, the shoot apex appears to be a self-organizing region.

Further evidence that the shoot apex is a self-determining region, and that it behaves as an organized whole, is provided by the observation that when the apex of *Lupinus alba* is divided into four sectors by vertical radial cuts through the centre, each of these sectors regenerates into a normal apex.

Other experiments have included the surgical isolation of the apex from surrounding tissues by four vertical tangential incisions, so that it stands on a plug of parenchymatous tissue. Under these conditions, in which the influence of the vascular tissue of the older parts of the shoot was removed, the apices of both *Lupinus* and the fern, *Dryopteris*, continued to behave in a perfectly normal manner and produced new leaf primordia in the normal phyllotactic sequence, indicating, once again, the self-determining properties of the shoot apex. Indeed, so far from being controlled by the acropetal development of pro-cambial tissue, there is much evidence that developing buds and leaves exert a stimulatory effect on the differentiation of vascular strands in the stem tissue below. For example, if young leaf primordia are removed from a shoot apex at a very young stage, vascular tissue fails to develop in the stem or, in the case of ferns, may be greatly modified or reduced. On the other hand, in callus cultures of chicory or lilac there is no differentiated vascular tissue, but if a bud is grafted into the callus, then vascular strands develop in the callus below the bud (p. 158) (Fig. 25B).

The stimulatory effect of a bud on the development of vascular tissue appears to be due to the hormones, especially auxins and gibberellins, which it produces (p. 110). There is also evidence that hormones play an important role in the regeneration of vascular tissue (p. 157).

From these various types of evidence it is apparent that in respect of many of its activities the vegetative shoot apex is a self-determining region. On the other hand, when flowers are initiated, so that a vegetative shoot apex is con-verted into a flowering one, it appears that in many species this transition occurs under the influence of a stimulus arising within the mature leaves in the older parts of the plant (p. 170).

ROOT APICES

The apical region of roots shows both similarities and differences in comparison with shoot apices. No lateral organs such as leaves are initiated at the root apex,

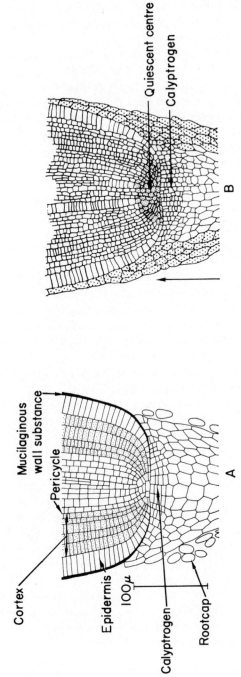

FIG. 26A. Diagram of longitudinal section of the root apex of a grass (*Stipa*) showing how files of cells can be traced back to a limited number of initials, and how the main zones of the root (epidermis, cortex and vascular cylinder) may be recognized at an early stage. (The calyptrogen is the meristematic zone which gives rise to the root cap.) (From K. Esau, *Anatomy of Seed Plants*, John Wiley & Sons, Inc., New York, 1960.)

B. Longitudinal section of the root of maize (*Zea mays*), showing the position of the quiescent zone (From F. A. L. Clowes and B. E. Juniper, *Plant Cells*, Blackwell Scientific Publications, Oxford, 1968.)

and hence growth is more uniform and there is no division into nodes and internodes. On the other hand, the apex of the root is covered by a root cap, which is not represented in the shoot apex.

It is possible, by careful study of the patterns of division in the root apical region, to trace back the origin of the main zones of the differentiated root, the epidermis, cortex and vascular cylinder, to certain groups of initial cells in the main zone of cell division, the *promeristem* (Fig. 26A). It has proved remarkably difficult to identify the initial cells with certainty, however, and there is still considerable difference of opinion as to the number of initial cells in roots. Some authors have claimed that there are relatively few initial cells—perhaps only three, or even one. On the other hand, Clowes has produced evidence that, in the root tip of *Zea mays* and other species, there is a group of rather inert cells which constitute the *quiescent centre* (Fig. 26B) and that the actively dividing initial cells occur at the boundary or "surface" of the quiescent centre. The zone of actively dividing cells takes the form of an inverted "cup". Several techniques have been used to study this problem, including autoradiographic studies to show the zones of active DNA synthesis and cell division, using a radioactive DNA precursor, such as ^{3}H-thymidine. Nuclei which show active DNA synthesis will incorporate the ^{3}H-thymidine, whereas inert cells will show no incorporation. In this way, the existence of the quiescent centre can be clearly shown in various species. The function of the quiescent centre is still obscure.

The initial cells and their immediate daughter cells are non-vacuolated and cell division proceeds actively in this zone. Further back along the root, however, cell division becomes less frequent and cell vacuolation and extension commence. In the roots of many species (e.g. wheat) growth is fairly sharply separated into regions of cell division and cell extension, but in others (e.g. beech, *Fagus sylvatica*) there is a certain amount of division in cells which are beginning to vacuolate.

The boundaries of the future vascular cylinder, cortex and epidermis of the root become recognizable at a short distance back from the apical initials. These zones can be distinguished by the sizes of the cells and by the planes of division; the cells of the inner layers of the cortex tend to develop by periclinal divisions, whereas the planes of the cell walls of the future vascular tissue are less regular.

The phloem procambium becomes recognizable at an early stage, by virtue of its small cells as seen in transverse section. The procambium develops acropetally and the differentiation of xylem and phloem follows in the same direction, the phloem preceding the xylem (Fig. 1).

It was shown earlier that, in many respects, the shoot apex is a self-determining region, and this appears to be true also for the root apex. Thus, the pattern of vascular differentiation appears to be controlled by the apex itself. For example, if the apical 2 mm of roots is cut off, and the tip is turned about its longitudinal axis and replaced on the stump, the vascular tissue which later differentiates in the tip is out of line with that of the stump. Again, Torrey cut off the extreme tips of roots of *Pisum sativum* and grew them in a suitable culture medium. It was found that whereas the original roots showed triarch xylem (i.e. it showed three protoxylem groups), a certain proportion of the regenerated roots showed a diarch structure. Thus, the experimental treatment destroyed the original pattern of differentiation and yet a new one was determined by the meristem.

FLOWER INITIATION AND DEVELOPMENT

Sooner or later, one or more of the vegetative apices of a plant cease to produce leaves and buds and become converted into flowering apices. This transition involves a basic change in the structure of the shoot apex. The first detectable change is an increase in cell division in the region of the corpus between the central mother zone and the rib meristem, and this increased cell division gradually spreads to the central mother zone and downwards into the flank regions. At the same time there is a marked reduction in cell division and growth in the rib meristem and pith where some of the cells undergo vacuolation. As a result of these changes, the shoot apical region is transformed into a structure consisting of a central pith of vacuolated cells covered by a "mantle" of smaller, densely staining meristematic cells (Fig. 27). During these changes the height of the apical region increases considerably in most species, but where the inflorescence is a capitulum, as in the Compositae, it may become flattened.

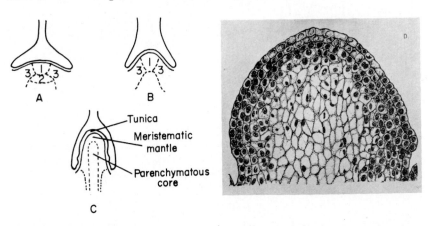

Fig. 27A, B, C. Modifications of zoning at shoot apex of *Succisa pratensis* during transition from a vegetative apex to an inflorescence: A. Vegetative apex; B. Early stage of transition; C. Later stage showing formation of meristematic "mantle", 1, Central mother cells, 2, Rib meristem, 3, Peripheral meristem; (the innermost sides of the enclosing leaf primordia are indicated as a single line above the apex) ; D. Early reproductive apex of *Xanthium*, showing the meristematic "mantle" overlying the enlarged central rib meristem. (From F. B. Salisbury, *The Flowering Process*, Pergamon Press, Oxford, 1963.)

The mantle, which includes both the tunica and the outer layers of the corpus, gives rise to the bracts and the flower primordia. Ultimately the flower primordia extend over the whole surface of the apex, so that all the meristematic tissue becomes differentiated. Thus, the structure of a vegetative apex becomes obliterated by the transition to a reproductive apex, and we have a change from an apex capable of unlimited growth to the determinate meristem of the inflorescence.

The subsequent pattern of development of the individual flower varies considerably from species to species. According to the "classical" viewpoint, the receptacle of the flower is a modified vegetative shoot and differs from the latter in that it is no longer capable of unlimited growth and has very short "internodes". The development of a relatively "primitive" flower, such as that of a buttercup (*Ranunculus*) (Fig. 28), bears out this view, since we find that during the early stages the apex still retains essentially the same structure as that of the vegetative shoot. Moreover, the initiation and early development of the various parts (perianth, stamens, carpels) are very similar to those of leaves, although the patterns of development diverge later. The stamen arises as a small protuberance and as it enlarges it gradually assumes the four-lobed form of the mature anther. The filament arises late in development, as the flower opens.

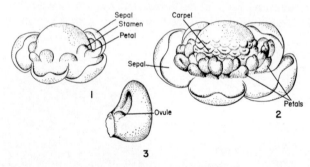

FIG. 28. Development of the flower in *Ranunculus trilobus*. 1 and 2, two stages in the development of the entire flower. 3. Developing carpel. (Adapted from Payer, *Traité d'organogennie comparée de la fleur*, Paris, 1857. Reprinted from A. Fahn, *Plant Anatomy*, Pergamon Press, 1967.)

In flowers with an apocarpous gynoecium, i.e. with free carpels, the first stage of carpel development is the appearance of a rounded primordium similar to that of the other organs. This primordium elongates and a depression appears in the tip. As a result of further unequal growth each carpel adopts a horseshoe form (Fig. 28); these structures grow upwards and their margins meet and fuse. In flowers with a syncarpous gynoecium, the carpels may arise independently at first and fuse later, or they may be already joined from the earliest stages.

FURTHER READING

General

CLOWES, F. *Apical Meristems*, Blackwell Scientific Publications, Oxford, 1961.

ESAU, K. *Plant Anatomy*, Wiley, New York; Chapman & Hall, London, 1963.

ESAU, K. *Anatomy of Seed Plants*, J. Wiley & Sons, Inc., New York and London, 1961.

FAHN, A. *Plant Anatomy*, Pergamon Press, Oxford, 1967.

LAETSCH, W. M. and R. E. CLELAND, *Papers on Plant Growth and Development*, Little, Brown & Co., Boston, 1967.

SINNOTT, E. W. *Plant Morphogenesis*, McGraw Hill, New York, 1960.

STEWARD, F. C. *Growth and Organization in Plants*, Addison-Wesley, Reading, Mass., 1968.

TORREY, J. G. *Development in Flowering Plants*, MacMillan, New York and London, 1967.

WARDLAW, C. W. *Morphogenesis in Plants*, Methuen & Co. Ltd., London, 1968.

More Advanced Reading

ALLSOPP, A. Shoot morphogenesis. *Ann. Rev. Plant Physiol.* **15,** 1964.

CUTTER, E. G. Recent experimental studies of the shoot apex and shoot morphogenesis. *Bot. Rev.* **31,** 7, 1965.

STEEVES, T. A. On the determination of leaf primordia in fern leaves, in *Trends in Plant Morphogenesis* (Ed. E. G. Cutter), Longmans, Green & Co. Ltd., London, 1966.

CHAPTER 3

General Aspects of Differentiation

ONE of the great problems associated with the study of differentiation concerns the way in which genetically identical cells give rise to the wide variety of cells, tissues and organs which make up the body of the mature plant. It is easy to show that the differentiated living cells of the plant body still retain the full genetical potentialities of the species, by experiments involving regeneration. Thus leaves of *Begonia* may be induced to regenerate whole plantlets from differentiated cells. Evidently, therefore, the cells of the leaf contain the genes involved in stem and root development, but they remain unexpressed in leaf tissue. This ability of differentiated cells of particular organs to regenerate a whole plant is described as "totipotency". The phenomenon of totipotency is demonstrated even more strikingly by certain tissue culture experiments in which it has been shown that single cells of carrot or tobacco cultures (originally derived from carrot root tissue or tobacco stem pith) may regenerate complete plants (p. 156). From these considerations we may draw the conclusion that probably all living cells of the plant body retain the full genetic potentialities of the original zygote. Thus, the manifold differences between the various cells of the plant body are evidently not due to differences in the genotype arising during differentiation. Rather, the problem is one of regulation of gene activity, i.e. why certain genes are active at certain phases of development, and other genes at other times. The control of gene activity is of intense interest to biologists and biochemists at the present time and we shall discuss this problem in more detail in a later chapter. However, some important factors in differentiation have been revealed by studies in experimental morphology, and a brief account of the results of such work will be given in the present chapter.

POLARITY

It is self-evident that plant species show characteristic form and one aspect of this form is that there is typically a well-developed longitudinal *axis*, bearing lateral organs such as leaves and flowers. Differences occur along the axis, so

48

that the two ends are not the same—for example, the plant axis is usually differentiated into a shoot end and a root end. In this respect the axis is said to show *polarity*, which has been defined as "any situation where two ends or surfaces in a living system are different". Polarity of the axis of plants is most readily seen from morphological differences, but we shall see that there may be physiological differences between the two ends of an organ even though morphological differences may not be apparent; indeed, morphological differences probably arise in the first instance from physiological differentiation.

Although axial polarity is one of the most striking features of the plant body, it should not be forgotten that there are other forms of polarity. For example, the *dorsiventrality* of leaves involves differences between the upper and lower sides and may be regarded as an instance of polarity. There may also be *radial polarity* in spherical bodies, such as cells of *Chlorella* or apple fruits, where there is a degree of radial symmetry, but there are differences between the inner and outer layers with respect to both chemical constituents and structure. In the following discussion we shall be concerned almost entirely with axial polarity and the term "polarity" will be taken to refer to this type, unless it is stated otherwise.

The Determination of Polarity

As we have seen, the root end of the embryo is directed towards the micropyle of the parent ovule and hence the orientation of the axis of polarity in embryos appears to be determined by conditions prevailing in the embryo sac, i.e. polarity is determined by the polarity of the tissues of the parent plant.

However, the polarity of the zygote is not predetermined in lower plants, such as the alga *Fucus*, the eggs of which are initially unpolarized. The spherical fertilized egg soon shows polarity, however, since one side begins to grow out as the future rhizoidal end, even before the first cell division has occurred. It can be shown that the polarity of the fertilized *Fucus* egg may be induced by light; thus, if the egg is exposed to unequal illumination on opposite sides, then the less strongly illuminated side becomes the future rhizoidal end. Light has a similar effect in determining polarity in spores of *Equisetum* (Fig. 29A), where the orientation of the first cell wall is at right angles to the direction of the incident light.

Other factors which may influence the induction of polarity in eggs and spores include gradients of pH, CO_2 and O_2.

The Persistence of Polarity

Once polarity has been induced in an organism, whether an embryo or a flowering plant, a fertilized *Fucus* egg or a spore of *Equisetum*, it becomes

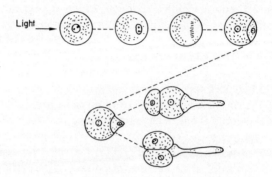

FIG. 29A. Induction of polarity in a spore of *Equisetum* by unilateral illumination. (From D. von Wettstein, *Encycl. Plant Physiol.*, **15** (1), 1965.)

FIG. 29B. Single cell from a filament of *Cladophora* regenerating a thallus from its apical end, and a rhizoid from the basal end. (Redrawn from A. T. Czaja, *Protoplasma*, **11,** 601, 1930.)

extremely difficult or impossible to reverse it. This fact is well illustrated in regeneration experiments with flowering plants. Thus, if we take stem pieces of a plant such as willow and suspend them in a moist atmosphere, they will develop adventitious roots towards the morphologically lower end and the buds will tend to grow out most strongly at the upper end (Fig. 30). Similarly, if we take segments of the roots of chicory, dandelion or dock, and plant them in moist soil or sand, roots will develop mainly from the morphologically lower (distal) end and buds from the upper (proximal) end (Fig. 31). Thus, although the willow cutting, and still less the root cutting, does not show any

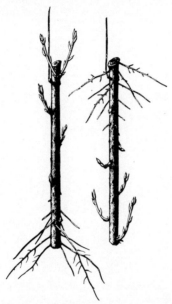

FIG. 30. Polarity of regeneration in willow stem. *Left*: stem cutting suspended in moist air in its normal orientation. *Right*: stem cutting similarly treated but in inverted position. Roots grew out at the morphologically lower end and shoot buds at the morphologically upper end, regardless of orientation. (Reprinted from *Pfeffer's Physiology of Plants*, 2nd edition, Clarendon Press, Oxford, 1903.)

morphological differentiation between the upper and lower ends, nevertheless it is clear that these organs possess a marked physiological polarity which affects the pattern of regeneration of roots and buds. This polarity is inherent in the tissues themselves and is not dependent on gravity, illumination or other external conditions, as is shown by the fact that willow cuttings suspended "upside down" in a moist atmosphere still regenerate roots predominantly at the original, morphologically lower end (Fig. 30), and similar effects are obtained with root segments of chicory which are planted so that the original upper end is now lowermost.

The physiological polarity of pieces of stem and root is a "built-in" property of the tissues. It might be thought that physiological polarity is due to gradients of metabolites or other substances along the stem or root, but this cannot be so, since polarity persists from one season to the next, through the dormant period, when metabolism is proceeding at a low rate, so that concentration gradients of metabolites are unlikely to persist. It is likely that the rooting of cuttings involves a gradient of growth hormone within the stem (p. 159), but this hormone gradient is one of the *results* of the polarity of the tissues and not the cause of it.

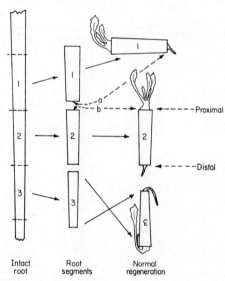

Fig. 31. Polarity of regeneration in root segments, such as those of *Taraxacum* and *Cichorium*. Shoot buds develop at the proximal end (i.e. the originally furthest from the root tip), and roots at the distal end, regardless of orientation. (Reprinted from H. E. Warmke and G. L. Warmke, *Amer. J. Bot.* **37**, 272, 1950.)

Cell Polarity

We have seen that a piece of stem shows polarity of regeneration, with a tendency for buds to develop from the upper end and roots from the basal end. If such a stem is divided into two, each half behaves in a similar manner, and this process can be repeated even with very short pieces of stem. If it were possible to carry out this process to the limit, we might conclude that each individual cell exhibits polarity, and indeed there is evidence that this is the case. Thus, the filamentous green alga, *Cladophora*, shows polarity of the plant body, in that the basal end forms a rhizoid. If the cells of a filament of *Cladophora* are plasmolyzed, so that the protoplasts are pulled away from the cell walls (which will break protoplasmic connections between cells), and are then deplasmo-lyzed, each cell subsequently regenerates a new filament, developing a rhizoid at the basal end of the cell (Fig. 29B). This experiment seems to provide clear evidence for polarity of individual cells of *Cladophora*. Comparable evidence for the cells of higher plants is more difficult to obtain, but much indirect evidence supports the hypothesis, such as the occurrence of unequal cell divisions, which will be described later (p. 54).

The Structural Basis of Polarity

Evidence has been presented which seems to suggest that (1) polarity of the tissues of a stem or root is remarkably stable and is not easily reversible, (2) polarity persists even throughout dormant periods, when metabolism is proceeding at a low rate, and (3) the polarity of a tissue apparently reflects the polarity of the individual cells. On these grounds, it has been suggested that there must be some permanent, structural basis for cell polarity. The observation that a piece of stem can be divided into several pieces, each of which shows the same polarity of regeneration, is very reminiscent of the fact that a bar magnet can be similarly divided and each piece becomes a small magnet. In a magnet each iron atom possesses magnetic polarity and the atoms become aligned during the process of magnetization, so that the "North" and "South" pole of each atom becomes oriented along the axis of the bar. By analogy, we might postulate that each cell of polarized organs, such as stems, contains polarized molecules (possibly elongated molecules with chemically different groups at the two ends) which are oriented along the axis of each cell to form a "cytoskeleton". Since the orientation of these molecules is not destroyed by the cyclosis of the cytoplasm in the cell, it would appear that the "cytoskeleton" is located in the boundary layers of the cytoplasm, where such cyclosis probably does not occur. It has been suggested that the cytoskeleton consists of longi-tudinally oriented fibrous protein molecules, but, attempts to demonstrate such oriented protein molecules in the cell have so far proved unsuccessful. It is interesting that the only linearly oriented structures that we do know of in the boundary layers of the cytoplasm, the microtubules, lie at *right angles* to the axis of polarization in stems and roots (p. 12).

POLARIZED CELL DIVISION

The planes of cell division during the development of an organ play a very important role in determining its final form and shape. Indeed, we may say that without oriented cell divisions there could be no organized form within the plant body, as we see in cultures of callus tissue, where the plane of cell division is at random, and the resulting tissue forms a shapeless and structurally unorganized mass (p. 151). Thus, polarized cell divisions give the plant body its three-dimensional form. For example, in a developing internode the majority of cell divisions are oriented so that the mitotic spindle lies parallel to the axis of the internode. Consequently, the internode grows mainly in length and relatively much less in diameter. For gourd fruits of different shape it has been shown that in long, elongated fruits divisions in which the mitotic spindle is oriented parallel to the long axis are much more frequent than divisions in which the spindle is in other planes, whereas in round fruits divisions in one particular plane do not predominate.

It would appear that the axis of polarity of the whole organ has a strong influence on the plane of cell division, by affecting the orientation of the mitotic spindle. How the orientation of the spindle is controlled is not known, but we have seen that it appears to be determined by the band of microtubules which appears in the cytoplasm before prophase, in the plane of the future cell plate (p. 12).

Not only is polarity of importance in determining the form of the plant body but, as will be seen in the next section, it is of decisive importance for other aspects of differentiation, and indeed Bünning has stated that there is "no differentiation without polarity".

UNEQUAL CELL DIVISION

Another aspect of the importance of polarity in differentiation is seen in the occurrence of unequal cell divisions. We have already shown that the first division of the fertilized egg of angiosperms is usually an unequal one (p. 25), resulting from the polarization of the cytoplasm of the cell. In this instance, the unequal division is the first event in the life-history of the new plant, but unequal divisions are also frequently seen in later stages of differentiation.

In the root epidermis of certain grasses, including *Phleum pratense*, root hairs arise from daughter cells formed by the unequal division of certain epidermal cells. These cells have their long axis parallel to the root axis, and it can be seen that the cytoplasm of these cells is more dense at the apical end. Mitosis occurs and a transverse cell wall is formed in a position which gives rise to a small daughter cell with dense cytoplasm and a larger cell with less dense cytoplasm (Fig. 32). The root hair initials are normally formed only from the smaller cells.

A similar situation is seen in the development of the stomatal guard cells of certain monocotyledons. Here also, certain epidermal cells of the developing leaf show unequal division, and cut off a small cell with densely staining cytoplasm at one end, and a larger cell at the other (Fig. 33A). The smaller cell

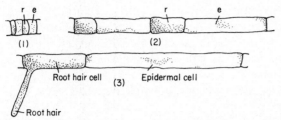

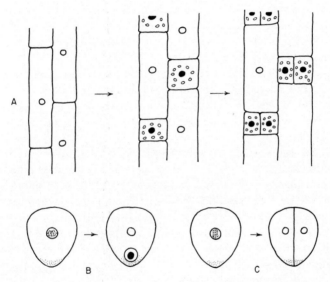

F<small>IG</small>. 32. Unequal division in the development of root hair initials in *Phleum*. 1, 2 and 3 are successive stages in development. Note that the last division (1) is unequal and the smaller cell (r) is destined to develop into a root hair cell, while the large cell (e) becomes an epidermal cell. (From E. W. Sinnott and R. Bloch, *Proc. Nat. Acad. Sci. U.S. 25,* 248–252, 1939)

F<small>IG</small>. 33A. Unequal division in formation of stomatal guard cells in leaf of a monocotyledonous plant. Epidermal cells undergo unequal division with the formation of a smaller stomatal mother cell (nucleus shaded) and a larger cell (nucleus not shaded). Chloroplasts develop in the stomatal mother cell, which divides again at right angles to the plane of the first division to produce two daughter cells which develop into the guard cells.

B, C. Development of pollen grain.

B. Normal development. Nucleus divides so that one of the daughter nuclei moves to the region of more dense cytoplasm at one end of the cell and becomes one generative nucleus. The other daughter cell nucleus becomes the vegetative nucleus.

C. Abnormal development. Nucleus divides at right angles to the normal plane. Daughter nuclei thus remain in the same cytoplasmic environment, and two equal cells are formed which disrupts the further normal development of the pollen grain. (Redrawn from E. Bünning, *Handbuch Protoplasmaforschung*, Vol. VII, Vienna, 1958.)

becomes a stomatal mother cell and it undergoes a further division, at right angles to the first division, and the two daughter cells so formed are in this case identical, and give rise to the guard cells. Thus, where a parent cell shows polarized differences in the cytoplasm along its axis, division is unequal and leads to differentiation between the resulting daughter cells, but where the plane of division appears to divide the cytoplasm equally, the resulting daughter cells are identical.

The fact that unequal division is preceded by polarized differences in the cytoplasm is well illustrated in the development of pollen grains in which the spindle is oriented so that one of the daughter nuclei passes to the end of the cell at which there is denser cytoplasm, and becomes the generative nucleus, whereas the other daughter nucleus moves to the region of the cell with less dense cytoplasm and becomes the vegetative nucleus (Fig. 33B). Occasionally the plane of the spindle accidentally becomes oriented *across* the axis of the pollen mother cell, and in this case two equal cells are formed and the further development of the pollen grain is disturbed (Fig. 33C).

It is not known how the polarized differences in the cytoplasm in the mother cell arise in the first instance, but the process of division itself is likely to lead to polarization, since the end of a daughter cell at which the equator was formed is likely to contain a different distribution of organelles from the end at which the pole of the spindle occurred.

In most of the examples we have just considered, the cells derived from unequal division do not themselves undergo further division, but differentiate directly. In some instances, however, the derivative cells from an unequal division undergo several more divisions. For example, in plants of the castor bean (*Ricinus communis*) there are secretory cells which contain tannin and unsaturated fatty acids. These cells rise from an initial unequal division, and one daughter cell undergoes a series of divisions to give rise to a row of cells, each of which becomes a secretory cell. Thus, the differentiated state can apparently be transmitted by cell lineage in some instances.

CELL LINEAGES IN DIFFERENTIATION

The question arises as to whether unequal division, followed by transmission of the resulting differences through further cell divisions (i.e. by cell lineage), are normally involved in differentiation in the shoot and root apices. When we consider certain lower plants, especially the algae and bryophytes, the importance of unequal division in differentiation is quite clear. For example, in the alga, *Chara*, the apical cell divides unequally, the outer daughter cell continuing as the apical cell and the other giving rise to differentiated thallus cells after a second unequal division (Fig. 34). In certain mosses and a few pteridophytes, such as certain species of *Selaginella*, the various tissues of the mature stem can apparently be traced back to precise divisions of the single apical cell and its immediate derivative cells, so that differentiation here also appears to arise by unequal division followed by transmission by cell lineage. However, in the shoots of other pteridophytes and of seed plants, it is not possible to trace any clear relation between the pattern of cell division at the shoot apex and the differentiation of the tissues derived from it.

In roots, whether those of pteridophytes with a single apical cell or those of seed plants, with no clearly identifiable initial cells, it is possible, by studying the patterns of cell division, to trace the cell lineages (i.e. groups of cells derived from a common parent cell) in the root tip region (p. 43, Fig. 26A). From such studies it appears that in many Dicotyledons, separate groups of initial cells give

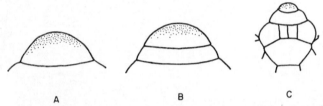

A   B   C

FIG. 34. Apex of *Chara*, showing unequal division of apical cell (A, B). Outer daughter cell continues as apical cell, and inner daughter cell gives rise to differentiated thallus cells after a second unequal division (C). (From E. Bünning, *Handbuch Protoplasmaforschung*, Vol. VII, Vienna, 1958.)

rise to the central cylinder, to the cortex, and to the apex and root cap, respectively. In some Monocotyledons, the central cylinder, the cortex and the root cap arise from separate initials and the epidermis is derived from the outermost layer of the cortex. However, we do not know whether the cells of any given lineage are already irrevocably *predestined* to give rise to certain tissues, as was held by earlier workers; that is to say, we do not know whether (1) the early derivatives of the initial cells of the promeristem are already differentiated from each other and these differences are then transmitted through further divisions to give rise to the various tissue zones of the mature root, or (2) the cells of all lineages are, at first, still capable of giving rise to any of the main zones of the root, and only later is the pattern of differentiation imposed on them by some unknown processes. However, even if the former type of mechanism holds for the roots, it is difficult to see how such a mechanism could be applied to shoots where the origin of organs, such as leaves, or of vascular tissue, cannot be directly related to specific cell lineages in the apical region.

THE INDUCTION OF SIMILAR DIFFERENTIATION
PATTERNS IN NEIGHBOURING CELLS

So far we have considered some of the factors which lead to *different* patterns of differentiation in neighbouring cells, but in the development of certain tissue systems it is necessary that similar differentiation should occur in adjacent cells. A good example is provided by vascular tissue, where the formation of a xylem vessel requires the co-ordinated differentiation of contiguous cells over a long distance in the trunks of many trees.

The most effective way of achieving continuity in the pattern of differentiation in a longitudinal chain of cells would seem to be for one differentiating vessel element to induce similar differentiation in the next adjacent cell. Evidence for such *homeogenetic induction* of differentiation by one developing vascular cell on another is seen in the regeneration of vascular tissue in the stem of *Coleus*.

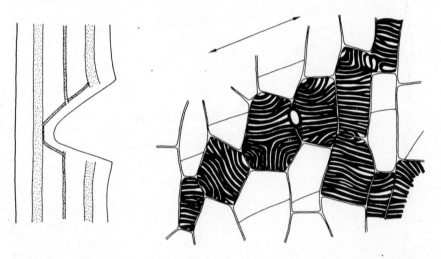

Fig. 35. Vascular regeneration in *Coleus*. *Left:* regeneration of connection between several vascular bundles in stem. *Right:* differentiation of parenchyma cells into reticulate xylem cells in the development of this strand. Arrow shows the direction of its development. (From E. W. Sinnott and R. Bloch, *Amer. J. Bot.* **32,** 151, 1945.)

If one of the vascular bundles of this stem is severed by a lateral cut, regeneration of the strand occurs through the pith, and connects the severed upper and lower ends of the bundle (Fig. 35). The pattern of regeneration is basipetal, i.e. it starts from the upper severed end of the original vascular bundle and travels from cell to cell diagonally through the pith. The re-generated strand is formed by the differentiation of parenchymatous pith cells into xylem and phloem. The xylem elements are formed by the development of reticulate lignified thickenings on the cell walls. During the early stages of differentiation of these cells granular strands (shown to contain microtubules which lie parallel to the future thickenings in the wall), appear in the surface layers of the cytoplasm and mark out the position of the lignified bands which later develop on the cell walls. Moreover, the thickenings in one cell are opposite those of the next, so that pattern can be observed in adjacent cells which extend across the cell boundaries.

These observations strongly suggest that the pattern of differentiation in one cell may induce similar changes in an adjacent cell. How this homeogenetic induction of differentiation is achieved is not known, but the hormone indole-acetic acid appears to be an important factor in xylem regeneration and differentiation. In *Coleus* it has been shown that the presence of a leaf above the cut in the stem is essential for the regeneration of the strand, but that the leaf can be replaced by applying auxin to the petiole stump. This result also suggests that auxin is necessary for the regeneration of vascular tissue. However, it is questionable whether auxin is the only factor involved, since it stimulates the regeneration of both xylem and phloem tissue, so that its effect would seem to be relatively non-specific. Moreover, the strict localization of differentiation to well-defined strands of cells, and the extension of cell wall patterns across the boundaries of adjacent cells, suggest that other factors, in addition to auxin, are involved in the homeogenetic induction of vascular tissue.

PATTERN IN DIFFERENTIATION

Patterns in the distribution of differentiated cells and tissues are very common and can take various forms. Thus, patterns may be seen in the arrangement of root hairs in regular longitudinal rows or in the regular markings seen in the petals of many flowers. Another example of pattern is uniform distribution of structures, such as leaf hairs, over a surface so that they are separated from each other at approximately the same distance, forming a "mosaic". Other forms of pattern are manifold.

One factor which plays an important role in pattern formation is unequal division, as we have seen. Another phenomenon which seems to underlie pattern formation is the *mutual inhibition of like structures*. An example is seen in the distribution of stomata in a developing leaf (Fig. 36). We find that the first series of stomata are uniformly distributed throughout the surface of the young leaf, but as this expands and the first formed stomata become separated by a greater distance, then new guard mother cells arise in the spaces between the original stomata. This pattern of behaviour suggests that developing structures of like nature, in this case guard mother cells, exert a mutually inhibitory influence on each other, so that when one structure occupies a given position, no similar structure can arise within a certain minimum distance from it. The cause of this inhibition between like structures is unknown but it has been suggested that it is due to competition between developing structures for specific substances required for their differentiation, so that if one structure arises at a given point it will tend to monopolize these substances within a certain radius, and so will prevent a second similar structure arising within its "inhibitory field".

Very often isolated cells, such as guard mother cells show cell division after the surrounding cells have ceased to divide and Bünning has called them *meristemoids*. He suggests that mutual inhibition is a property of such meristemoids and he has shown that a number of other examples of pattern can be

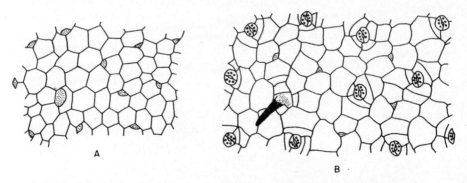

A

B

FIG. 36. Mutual inhibition between guard mother cells. Patterns of stomatal mother cells in a dicotyledonous leaf. A. The initial cells are shaded, but the large cell is a hair initial. B. Later stage in which, due to expansion of the leaf, these initial cells have moved further apart and developed into guard cells. The enlarged spaces between these allow the formation of new stomatal initials. (Redrawn from E. Bünning, *Survey of Biol. Progress*, Vol. 2, p. 105, Academic Press, New York, 1952.)

interpreted in terms of this hypothesis. For example, he suggests that the formation of discrete strands rather than a continuous cylinder, of procambial tissue in the stem, is due to mutual inhibition of like structures, so that a strand can only develop at a certain minimum distance from a neighbouring strand. The origin of leaf primordia at the shoot apex may provide a further example of such inhibition (p. 33).

CELL DIFFERENTIATION AND ENVIRONMENTAL FACTORS

So far we have considered primarily the "internal" factors, arising within the plant itself, in differentiation. However, it is well known that environmental factors may also have a profound effect on the pattern of differentiation and, indeed, Chapters 9 and 10 are largely concerned with the effects of environmental factors, especially daylength and temperature, on the development of the plant as a whole, including the transition from the vegetative to the reproductive phase. Apart from this environmental control of a major phase-switch in the development of the whole plant, physical factors, such as light, temperature, water supply, oxygen tension, mineral nutrient level and gravity, are all known to have effects on tissue and cell differentiation. The profound effects of light on the form and structure of the shoot will be described in connection with photomorphogenesis (p. 179).

Apart from major switches in differentiation of this type, however, environmental factors have very widespread effects as modifying factors. For example, plants grown under conditions of water-stress are frequently found to show more pronounced lignification of tissues than those grown under more favourable water regimes. Another example is provided by the occurrence of "sun" and "shade" leaves in woody plants; the leaves of trees exposed to full sunlight have thicker palisade layers than those growing on the north side or in shaded conditions on the lower branches. Differentiation takes place earlier in shade leaves than in sun leaves.

We have seen that submergence in water has a marked effect on leaf shape in certain aquatic plants (p. 39), and it also affects the anatomical structure of some species. Thus, in certain species the development of *aerenchyma*, i.e. tissue with well developed intercellular spaces, is developed only in those parts of the stem or root which are submerged, so that there is a sharp transition in structure at the water–air boundary. Apparently the important factor in this phenomenon is oxygen tension, which is much lower in water than in air.

FURTHER READING

General

BONNER, J. T. *Morphogenesis*, Atheneum, New York, 1963.
BÜNNING, E. Morphogenesis in plants. *Survey of Biological Progress*, **2,** 105, 1957.
SINNOTT, E. W. *Plant Morphogenesis*, McGraw Hill, New York, 1960.
SINNOTT, E. W. *The Problem of Organic Form*, Yale Univ. Press, New Haven and London, 1963.

More Advanced Reading

BLOCH, R. Polarity in plants, *Bot. Rev.* **9,** 261, 1943.
BLOCH, R. Histological foundations of differentiation and development in plants. *Encycl. Plant Physiol.* **15**(1), 146, 1965.
BLOCH, R. Polarity and gradients in plants: a survey. *Encycl. Plant Physiol.* **15**(1), 234, 1965.
CUTTER, E. G. (Ed.). *Trends in Plant Morphogenesis*, Longmans, Green & Co. Ltd., London, 1966.
LANG, A. Progressiveness and contagiousness in plant differentiation and development. *Encycl. Plant Physiol.* **15**(1), 409, 1965.

CHAPTER 4

Plant Growth Hormones: Their Discovery and Chemistry

THE growth of a plant is a dynamic and complex, yet strictly controlled, process. This means that growth in different parts of the plant must be integrated and co-ordinated and we shall meet a considerable number of examples of such *growth correlations* in later chapters. The co-ordination of growth between different parts of the plant must clearly involve some control mechanism. Moreover, we have seen, in earlier chapters, that the development of organs, such as leaves or stems, involves an orderly sequence of phases of cell division and cell extension, so that there is also co-ordination of growth in *time*. As a result of intensive studies extending over many years, it is now known that hormones play a vital role in the control of growth, not only within the plant as a whole, but apparently also within individual organs. It is now realized that there are at least three major classes of growth-promoting hormones—*auxins*, *gibberellins* and *cytokinins*. In addition, other classes of plant hormones exist, particularly the "growth inhibitors" such as *abscisic acid* (ABA) (see Chapter 11), but also including a gas, *ethylene*, which is apparently involved in many growth phenomena.

Growth hormones are translocated within the plant, and influence the growth and differentiation of the tissues and organs with which they come into contact. This leads us to a consideration of the nature and role of growth hormones. The word "hormone" was first used by animal physiologists, to refer to a substance which is synthesized in a particular secretory gland and which is transferred in the blood or lymph to another part of the body where extremely small amounts of it influence a specific physiological process. However, plant hormones differ in certain respects from the classical concept of hormones which was originally based upon the discovery of these substances in animals. In an animal a hormone is a substance produced in one particular organ such as a gland, and which is secreted to produce its typical and usually specific effect at a site distant from its point of origin. In the case of plant hormones we cannot always differentiate so clearly between the site of hormone synthesis and its place of action, although there is much evidence, referred to in Chapters 5 and 6, that plant hormones do usually have effects at sites distant from their place of production. Another difference between animal and plant hormones is that whereas the effects of most animal hormones are rather specific, a plant hormone can elicit a wide range of

62

responses depending upon the type of organ or tissue in which it is acting. For reasons such as these, plant growth hormones have frequently been referred to by other names, such as "growth regulators" or "growth substances". In general, though, the term "growth hormone" appears to be more appropriate, despite the difficulties discussed above, since it does intimate that these substances are active in extremely small quantities, and that they exert control over processes in tissues different from those in which they are synthesized.

As mentioned earlier, there are three types of growth-promoting hormone known to be operating in plants, auxin, gibberellin and cytokinin. Each of these chemically different types of growth hormone has characteristic influences on the growth and differentiation of plant cells. Auxins were the first plant growth hormones to be discovered, and consequently we shall now review very briefly the history of the discovery of auxins and their chemistry, followed by similar considerations of gibberellins, cytokinins and finally ethylene. At the present time, abscisic acid is more appropriately considered in connection with dormancy and photoperiodism in plants (see Chapter 11).

AUXINS

The basis of our modern knowledge of auxins lies in the work of Charles Darwin, published almost a century ago in a book entitled *The Power of Movement in Plants*. Darwin investigated the phenomenon of *phototropism*, the bending of plant organs in response to unilateral illumination.

Darwin experimented with seedlings of the ornamental canary grass (*Phalaris canariensis*). The coleoptile of grass seedlings proved a very convenient subject for the study of phototropism by Darwin and many other later workers. However, it was Darwin who first realized that the tip of the coleoptile *perceives* the unilateral light stimulus, but that the curvature *response* occurs lower down (Fig. 37). Darwin concluded that, "when seedlings are freely exposed to a lateral light some influence is transmitted from the upper to the lower part, causing the latter to bend". It was left to later researchers following Darwin to find out the nature of the "influence".

Various workers, particularly Boysen-Jensen and Paal, conducted experiments in the second two decades of this century which demonstrated that the growth-promoting influence transmitted from a coleoptile tip was of a purely chemical nature (Fig. 37). Paal was led to suggest that this chemical, under conditions of darkness or uniform illumination, continually moves down the coleoptile on all sides and acts as a *correlative growth promoter*.

The first successful isolation of the chemical messenger from coleoptile tips was carried out in 1926 by a Dutchman, F. W. Went, who thus extended the work of Boysen-Jensen and Paal. Went found that if he placed an excised oat (*Avena*) coleoptile tip upon a small block of agar gel, then the agar block

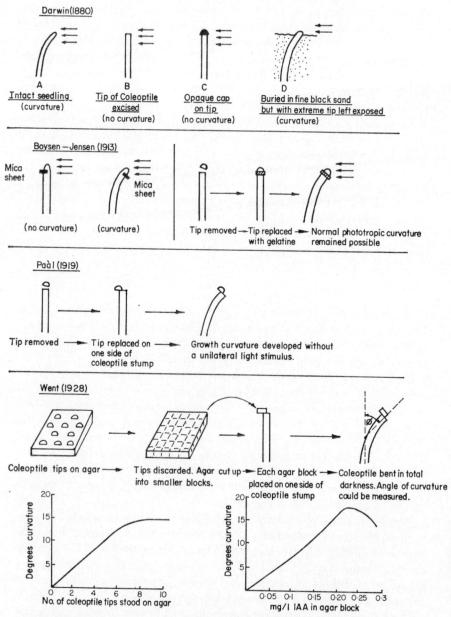

FIG. 37. A summary of important experiments which established the existence of auxin in plants. All experiments performed with coleoptiles of grass seedlings. Triple arrows indicate the direction of incident unilateral light. In the case of Went's experiments, dose-response curves are shown for the Went *Avena* curvature test; one for response against number of coleoptile tips diffused on agar, and the other for response against concentration of indole-3-acetic acid (IAA) in the agar block.

acquired growth-promoting properties, in that if the block was separated from the tip and placed on one side of a coleoptile stump, then curvature of the stump resulted. An agar block which had not been in contact with a tip had no such effect. The conclusion was, therefore, that the chemical messenger had diffused from the tip into the agar block. Subsequent placing of the agar block on to a coleoptile stump allowed diffusion of the messenger out of the block and into the stump. However, this was not only the first separation of a growth hormone from a plant but it also afforded Went a technique whereby he was able to make quantitative measurement of the growth hormone. The method of measurement that Went devised is a biological assay (*bioassay*) based on the curvature of a coleoptile stump in response to an asymmetric placing of an agar block containing the growth hormone (Fig. 37).

The name *auxin* (from the Greek *auxein*, to grow) was given to the growth hormone produced by the tip of a coleoptile. However, it is now known that auxins are present in *all* higher plants and not only in grass seedlings. Auxins are, as we shall see later, of fundamental importance in the physiology of growth and differentiation. They appear to be synthesized mainly in meristematic tissues such as those of the stem and root apex, young developing leaves, flowers and fruits.

The Isolation and Chemical Characterization of Auxin

Early attempts to isolate and chemically identify auxin suffered from the difficulty that it is present in plant tissues in extremely small quantities. It was found to be impossible at that time to obtain sufficient auxin from plants to prepare pure crystalline auxin for analysis. In fact the first crystalline auxin was obtained from human urine. Some confusion arose as a result of early attempts to determine the chemical nature of this auxin, but in 1934 it was shown to be indole-3-acetic acid (often abbreviated to IAA).

Indole-3-acetic acid (IAA)

Since the initial discovery of IAA as an auxin, it has been found that this substance occurs in most plant species, and it is now believed to be the principal auxin in higher plants.

Much research is still conducted with the aim of elucidating the true nature of endogenous (i.e. internally synthesized) auxin in plants, and this involves the use of a number of modern chemical and physical techniques, as well as biological methods. As we saw earlier, the first isolation of auxin from plants was achieved by allowing diffusion of the hormone to take place from the plant tissue into a suitable inert medium such as agar gel. This method is still utilized and auxin obtained in this way is referred to as *diffusible auxin*. More commonly, plant tissues are extracted by organic solvents, such as di-ethyl ether or methanol. There are often quantitative and qualitative differences between auxins obtained by diffusion and extraction, even from the same tissue.

Chromatographic separation of the constituents of a plant extract results in the original extract being split up into a number of separate fractions. Each of these fractions is then examined, to see whether or not it contains any auxin. To do this some sort of bioassay is usually used, such as the Went *Avena* test described earlier. There are numerous other types of biological assay for auxin that have been developed over the years, all of them making use of some measurable aspect of the effects of auxin in plants. One of the most commonly used tests is the coleoptile straight-growth test, in which sections of uniform length (say 5 mm) are cut from young *Avena* coleoptiles and placed in the extract to be tested for growth-promoting activity. However, although biological assays are indispensable in such studies, chemical identification of the auxin must eventually be carried out. Even today this identification presents great difficulties, mainly because of the difficulty in obtaining sufficient auxin for analysis. Nevertheless, several sensitive physico-chemical techniques, including ultraviolet and infrared spectroscopy, have been evolved which enable such identifications to be made.

The results of such studies have, in general, confirmed that IAA occurs as an auxin in plants, although it has become apparent that there are other substances in plants which also possess auxin activity. In many cases these other substances are indoles, closely related chemically to IAA; for example, the compound indole-3-acetonitrile (IAN) is known to be present in a number of plant species.

Indole-3-acetonitrile (IAN)

There are also compounds present in plants which are not indoles and yet possess auxin activity, but both the chemical natures and the physiological significance of these non-indole endogenous auxins remain obscure at the present time. It is because of the existence of more than one type of chemical which show auxin activity, that we often speak of auxins, rather than the singular, auxin.

Indolic Auxin Metabolism

Investigations of the metabolism of endogenous auxin have concentrated on the origin and fate of IAA in plant tissues. The reasons for this are first, the belief that IAA is the principal auxin, and secondly the concern to understand normal plant growth regulatory mechanisms. Since it is known that the concentration of auxin available to tissues can have a determining effect on growth and differentiation, then the factors which limit the concentration of IAA in plant tissues have received most attention. These factors include:

1. IAA synthesis.
2. IAA destruction.
3. IAA inactivation by processes other than destruction of the molecule.

Auxin biosynthesis. As soon as IAA was identified as an endogenous auxin, it was suggested that it is formed from the amino acid tryptophan, a compound with an indole nucleus which is universally present in plant tissues, both in the free state and incorporated into protein. Indeed, it has been shown that when exogenous tryptophan is supplied to higher plants under non-sterile conditions, they are able to convert it to IAA. However, under *sterile* conditions oat coleoptiles are unable to convert tryptophan to IAA, and their apparent ability to do so under non-sterile conditions is evidently due to the presence of contaminating bacteria which are able to carry out this conversion. Nevertheless, it is possible that some higher plants, such as pea, are able to convert tryptophan to IAA through indole pyruvic acid (IPyA) and indole-3-acetaldehyde (IAAld) (Fig. 38). In oat coleoptiles there is evidence that IAA may be formed from tryptamine, which in turn may be derived from indole. Further, there is no doubt that in some species indole-3- acetonitrile (IAN) can serve as a precursor of IAA, but the pathway of synthesis of IAN itself is still unknown. The pathway of biosynthesis of IAA is thus very uncertain at present.

Apart from the indole compounds shown in the schematic representation of IAA synthesis in Fig. 38, a number of other indoles are known to occur naturally in plants. It is possible that any one or all of these indoles could serve as precursors of IAA, but clearly we still lack good information on the biosynthetic pathways for IAA actually occurring *in vivo*.

IAA destruction. It is well established that IAA is in one way or another fairly readily inactivated by most plant tissues. It appears, therefore, that the concentration of IAA in plants is regulated not only by its rate of synthesis, but also by inactivation mechanisms. The indications are that IAA catabolism as well as biosynthesis may follow more than one pathway.

(a) *Photo-oxidation.* IAA in aqueous solution will soon decompose if left exposed to light. This photo-oxidation of IAA is greatly accelerated by the presence of many natural and synthetic pigments. Thus, it is possible that plant pigments absorb light energy which energizes the oxidation of IAA *in vivo*. If this is so, then the most likely pigments involved are riboflavin and violaxanthin,

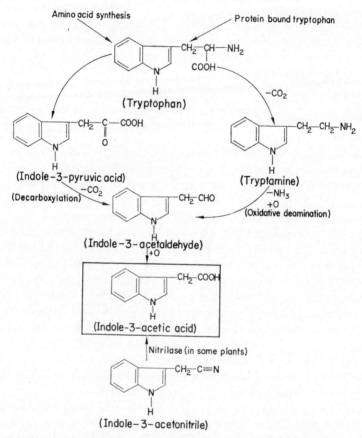

FIG. 38. Possible biosynthetic pathways for indole-3-acetic acid (IAA) in plants. (The pathway though tryptamine has been proposed, but the evidence for this pathway is not yet conclusive).

for these are commonly present in plants in relatively large amounts and they absorb light in the blue regions of the spectrum which have been found to be the most active in inducing photo-oxidation of IAA.

Breakdown products of *in vitro* photo-oxidation of IAA include 3-methylene-2-oxindole and indolealdehyde, together with other unidentified compounds formed by cleavage of the indole ring.

Indolealdehyde   3-methylene-2-oxindole

Nothing is known of the chemistry of IAA photo-oxidation within the plant, but indolealdehyde and methylene-oxindole do occur naturally in many plants, and may perhaps represent a product of *in vivo* photo-oxidation. Methylene-oxindole is markedly inhibitory to growth at some concentrations, and it is possible that the inhibition of growth by light is due to the formation of this compound in the tissues.

(b) *Enzymic oxidation of IAA*. Many plants contain an enzyme, or enzyme system, known as *"IAA-oxidase"*, which catalyses the breakdown of IAA, with the release of CO_2 and consumption of O_2. IAA-oxidase preparations from different plant species often have different properties, but they all show some similarities to the enzyme peroxidase. Full peroxidation of IAA (i.e. by H_2O_2 in the absence of O_2) does not occur, but oxygen is always required in addition to H_2O_2. Addition of H_2O_2 to some IAA-oxidase preparations does enhance the rate of IAA degradation, but in other cases it does not; probably because crude plant enzyme preparations can possess the ability to generate H_2O_2 which would be available for peroxidase action. IAA-oxidase from higher plants requires manganese as a co-factor, and its activity is increased by monophenols and reduced by ortho-diphenols.

The pathway of IAA breakdown by IAA-oxidase is ill-understood. A principal product is 3-methylene-oxindole, which is probably further metabolised through 3-methyloxindole as an intermediate. Should, however, cytochrome and cytochrome oxidase be present in addition to IAA-oxidase, then indole-3-aldehyde is the main product of reaction.

Interest in the enzymic oxidative destruction of IAA has been maintained by indications that such a process might be important in regulating auxin levels in plant tissues. Thus, there have been a number of reports that (1) IAA-oxidase activity rises with age of tissues, (2) there is a negative correlation between growth rate and IAA-oxidase content of various organs, and (3) that root tissues contain both very low IAA concentrations and very high IAA-oxidase activity. Nevertheless, there is not, as yet, conclusive proof that such correlations are physiologically important. Consequently a great deal of research is still being conducted to evaluate the role of IAA-oxidase in plant growth and differentiation.

Inactivation of IAA by other processes. Apart from degradation by photo-oxidation or enzyme activity, IAA can be inactivated in plant tissues by the formation of hormonally-inert complexes of various types. Thus, IAA is readily esterified by plant enzymes to yield its ethyl ester (indole ethyl acetate).

Indole ethyl acetate

Similarly, enzymic formation of conjugates such as indole acetyl-aspartic acid have been reported many times.

Indole acetyl-aspartic acid

The nature of the esterifying or conjugating enzymes involved in these reactions remains very uncertain at present.

Conjugates are also formed between IAA and various sugars and sugar-alcohols, yielding compounds such as IAA-arabinose, IAA-glucose and IAA-myoinositol, and between IAA and proteins.

GIBBERELLINS

The discovery of a group of plant growth-hormones now known as the gibberellins dates from the 1920's, when Kurosawa, a Japanese research worker, was investigating the "bakanae" ("foolish seedling") disease of rice caused by the fungus *Gibberella fujikuroi* (also known as *Fusarium moniliforme*). A characteristic symptom shown by rice plants infected by this fungus is an excessive elongation of stems and leaves, resulting in abnormally tall plants which usually fall over due to the spindly stem structure—hence, "foolish seedling". Kurosawa and fellow Japanese workers found that if they grew the fungus in a culture medium, and subsequently filtered off the fungus, then the culture filtrate, which was completely free of the fungus itself, would induce the same abnormal growth symptoms when applied to rice plants. It was clear that *Gibberella fujikuroi* secreted some substance into infected plants, or into the nutrient medium when grown in culture, which was stimulatory to stem and leaf elongation. In 1939 a small quantity of highly active crystalline material was isolated from such culture filtrates, and was given the name "gibberellin A". The chemical composition and structure of this material was not unequivocally worked out by the Japanese. It was not until 1954 that further progress was made, when British chemists isolated and chemically characterized a pure compound from culture filtrates of *Gibberella fujikuroi*. They called this new substance *gibberellic acid* and found that it has the following structure:

Gibberellic acid (GA$_3$)

It will be noticed that this structure is very different from that of auxins, such as indole-3-acetic acid.

Gibberellic acid when applied to many species of intact growing plants induced abnormally great extension of stems and leaves, but the response was found to be greatest when *genetic dwarfs* of various plant species were treated. Such treated dwarf plants assumed the appearance of the normal tall plants, from which the dwarfs had originally arisen by mutation (Fig. 39).

FIG. 39. Effect of gibberellic acid (GA$_3$) on shoot growth in dwarf pea plants (*Pisum sativum* c.v. Meteor). Plant at extreme right was not treated with GA$_3$, but the other plants received increasing doses of GA$_3$ from right to left. (Original print provided by Professor P. W. Brian.)

It might appear strange that a substance obtained from a fungus should produce essentially normal responses in higher plants. However, it is now known that gibberellic acid and substances very similar in both chemical structure and biological activity occur in healthy (i.e. non-infected) plants of all species. In fact, a number of these compounds have been isolated from higher plants, so that at the present state of knowledge there are approximately forty chemically characterized compounds which produce effects similar to those elicited by gibberellic acid. These compounds are known collectively as the gibberellins, designated gibberellins A_1, A_2, A_3, A_4, etc. Gibberellic acid is numbered A_3. Some of the known gibberellins have been isolated from culture filtrates of *Gibberella fujikuroi*, and others from various organs of higher plants. All have the same basic molecular structure (the gibbane carbon skeleton) as gibberellic acid, differing from one another mainly in the number and positions of substituent groups on the ring system, and the degree of saturation in the "A" ring (see Fig. 45). It is highly likely that further additions will in future years be added to the list of gibberellins.

The gibbane carbon skeleton

Apart from gibberellins having the gibbane carbon skeleton, there are some other compounds known which possess gibberellin-like biological activity but differ markedly in chemical structure. For example, a compound called helminthosporol isolated from the fungus *Helminothosporium sativum*, and phaseolic acid obtained from bean seeds.

The discovery that the gibberellins, or at least some of them, are natural growth hormones in higher plants, necessitated a complete reconsideration of views of the hormonal control of plant growth and differentiation. One could no longer think of development of cells and tissues as being influenced by only one growth hormone, that is auxin, but consideration had to be made of the effects of gibberellins and, later, other hormones (e.g. cytokinins) on the processes affected by auxins. Indeed, as we shall see later, auxins, gibberellins and cytokinins *interact* in their influences on plant growth and differentiation, and it is highly likely that abscisic acid and ethylene also interact with other growth hormones.

Gibberellin Metabolism

Gibberellins are chemically closely related to diterpenes, which are themselves members of a vast group of naturally occurring compounds in plants called terpenoids. Considerable knowledge of terpenoid biochemistry exists, and consequently rapid progress has been made in elucidating the outlines of gibberellin biosynthesis.

Gibberellin biosynthesis. All terpenoids are basically built up from "isoprene units", which are five-carbon (5-C) compounds. The linking together of two

$$\begin{array}{c} C \\ \diagdown \\ C \diagup \end{array} C - C - C$$

Isoprene unit (5-C)

isoprene units yields a monoterpene (C-10), of three a sesquiterprene (C-15), of four a diterpene (C-20), and of six a triterpene (C-30).

A summary of the biosynthetic sequence of gibberellins as we presently know it is shown in Fig. 40. This scheme has been shown to represent the pathway of gibberellin biosynthesis in the fungus *Gibberella fujikuroi*, but there is already evidence that a similar mechanism operates in higher plants.

It is of interest to note here that the growth inhibitor abscisic acid is a sesquiterpenoid, and it is possible that the initial steps in the biosynthesis of gibberellins and of this substance may involve a common pathway from mevalonic acid.

A number of synthetic *growth retardants* have been discovered in recent years, and some of these are proving of considerable importance in agriculture. Several of these growth retardants have been shown to act by inhibiting gibberellin biosynthesis in the plants to which they are applied. For example, the growth retardant "Amo-1618" (2-isopropyl-4-dimethylamine-5-methyl-phenyl-1-piperidine carboxylate methyl chloride) has been shown to inhibit the biosynthesis of gibberellin in homogenates of the endosperm of the wild cucumber (*Echinocytis macrocarpa*). It appears that Amo-1618 inhibits specifically the cyclization of geranylgeranyl pyrophosphate to (-)-Kaurene (Fig. 40). The growth retardant "CCC" ((2-chloroethyl) trimethyl-ammonium chloride) appears to act similarly.

Gibberellin catabolism. Very little is known of the eventual fate of gibberellins in plant tissues. There is evidence that gibberellins retain their physiological activity for some considerable time in plants. This is in marked contrast to the rapid inactivation of applied natural auxins (such as IAA) in plant tissues, but it is a common observation that even large doses of gibberellins are not particularly injurious to plants whereas auxins can be, perhaps due to an effect on the production of ethylene (p. 78). Thus, it would seem more important that plants should have the capacity for speedy inactivation of auxin when its concentra-

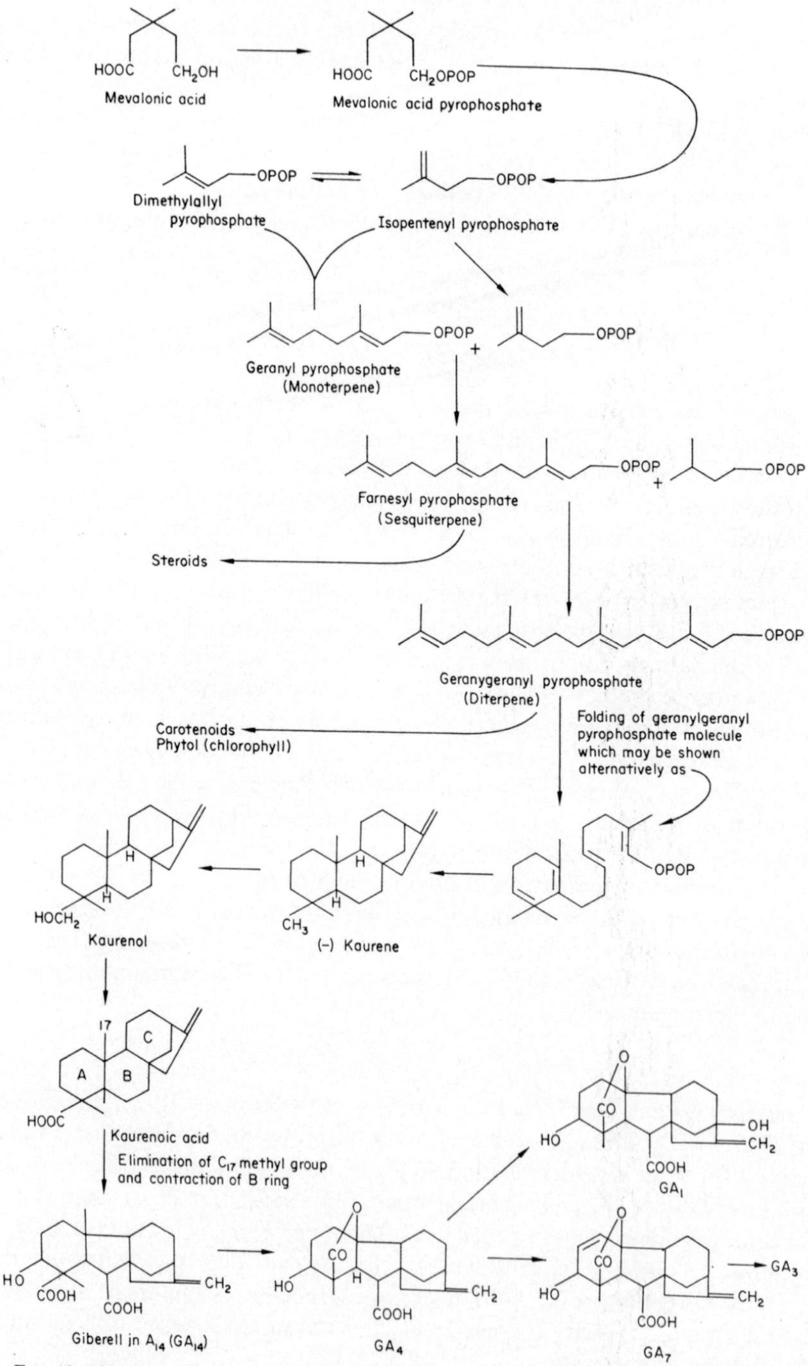

HOOC CH_2OH
Mevalonic acid

HOOC CH_2OPOP
Mevalonic acid pyrophosphate

OPOP
Dimethylallyl pyrophosphate

OPOP
Isopentenyl pyrophosphate

OPOP +
Geranyl pyrophosphate (Monoterpene)

OPOP
Farnesyl pyrophosphate (Sesquiterpene)

Steroids

OPOP +
OPOP

OPOP
Geranygeranyl pyrophosphate (Diterpene)

Carotenoids
Phytol (chlorophyll)

Folding of geranylgeranyl pyrophosphate molecule which may be shown alternatively as

OPOP

HOCH_2
Kaurenol

CH_3
(−) Kaurene

17
C
A B
HOOC
Kaurenoic acid

Elimination of C_17 methyl group and contraction of B ring

HO
COOH
COOH
CH_2
Giberell in A_14 (GA_14)

HO
CO
H
COOH
CH_2
GA_4

HO
CO
COOH
CH_2
GA_7

GA_3

HO
CO
COOH
OH
CH_2
GA_1

FIG. 40. The biosynthetic pathway for gibberellins from mevalonate. The sequence of reactions and transformations shown appear to hold in both fungi and higher plants.

tion exceeds a certain value. However, there is evidence that considerable *interconversion* of gibberellins takes place in plant tissues. That is, one particular gibberellin can be converted into a different gibberellin. Moreover, there is evidence for the existence of gibberellin glycosides (i.e. conjugates with sugars) in plant tissues, and these may represent the products of an inactivation mechanism.

Gibberellic acid in solution can be decomposed by acid hydrolysis, particularly at higher temperatures, to yield compounds such as gibberellenic acid, allogibberic acid and gibberic acid. The latter compound does not retain any hormonal activity, but gibberellenic and allogibberic acids can still elicit some of the physiological effects of gibberellins.

CYTOKININS

The discovery of this group of plant growth hormones came from work concerned with the *in vitro* culture of young plant embryos and tissue explants. Studies by Haberlandt in the first two decades of this century demonstrated the existence in plant tissues of a diffusible factor which stimulated parenchymatous cells in potato tubers to revert to a meristematic state. That is, cell division could be induced by the factor.

Many workers, particularly Skoog and Steward in the U.S.A., have made a close study of the growth requirements of callus-cultures (a mass of undifferentiated and usually rapidly dividing cells (p. 148)) prepared from the parenchymatous pith cells of tobacco, and from carrot roots. It is primarily as a result of their work during the 1950's that we became aware of the existence of cytokinins—plant growth hormones originally regarded as being particularly important in the processes of cell division and differentiation, but which more recently have been found to be implicated in various other physiological processes such as senescence and apical dominance.

Skoog used a tissue culture technique in which an isolated piece of tobacco pith is placed on the surface of agar gel into which had been incorporated various nutritive substances and other, hormonal, factors. The exact composition of the agar medium was varied, and the effects on the growth and differentiation of the pith cells noted. For growth to occur it was found that it was necessary to add not only nutrients to the agar, but also hormonal substances such as auxin. However, when auxin (IAA) was applied alone with the nutrients very little growth of the pith explant occurred, and that which did consisted predominantly of cell enlargement; very few cell divisions occurred, and no differentiation of cells took place. If, however, the purine base *adenine* was incorporated into the agar medium along with IAA, then the parenchymatous cells were induced to divide and a large callus mass was created. Adenine added without auxin, however, did not cause cell division in the pith tissue. There was, therefore, an *interaction* between adenine and auxin, resulting in the triggering off of cell division. Adenine is a purine derivative (6-aminopurine), and is a component of nucleic acids.

Later, another substance with the same, but more potent, effects as adenine was prepared from degraded deoxyribonucleic acid (DNA). This substance was found to be 6-(furfurylamino) purine, basically similar in structure, therefore, to adenine. Due to its property of actively promoting cell division (in conjunction with auxin) it was given the name of *kinetin*. Indole-3-acetic acid and kinetin were found to have interacting effects upon cell division and differentiation in tobacco pith cultures (Fig. 75), similar to those shown by IAA and adenine.

Adenine (6-amino purine)   Kinetin (6-furfurylamino purine)

Endogenous Cytokinins

Kinetin has never been shown to be present normally in plants, but substances which produce similar physiological and morphological effects have been found in various organs of many plant species, particularly in "nurse tissues" such as coconut milk (a liquid endosperm), in immature caryopses of *Zea mays*, and in immature fruits of *Aesculus hippocastanum* (horse chestnut), banana and apple. These naturally occurring substances, together with other synthetically prepared compounds which have effects on growth similar to those of kinetin, have been given the generic name of *cytokinins*. A cytokinin, therefore, is a substance which, in combination with auxin, stimulates cell division in plants and which interacts with auxin in determining the direction which differentiation of cells takes.

Most available evidence suggests that naturally occurring cytokinins are purine, particularly adenine, derivatives, In 1964 the New Zealander, Letham, isolated a cytokinin from sweet corn kernels and identified it as 6-(4-hydroxy-3-methyl but-2-enyl) amino purine. For convenience, Letham has called this substance *Zeatin*.

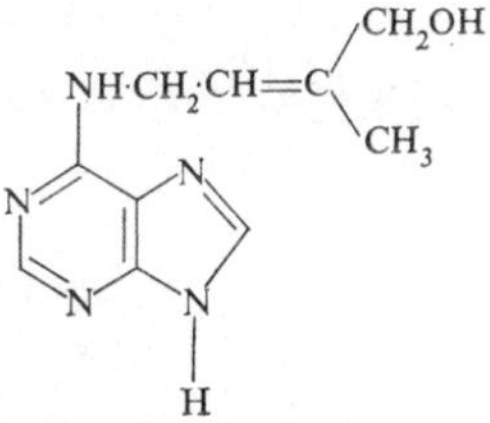

Zeatin

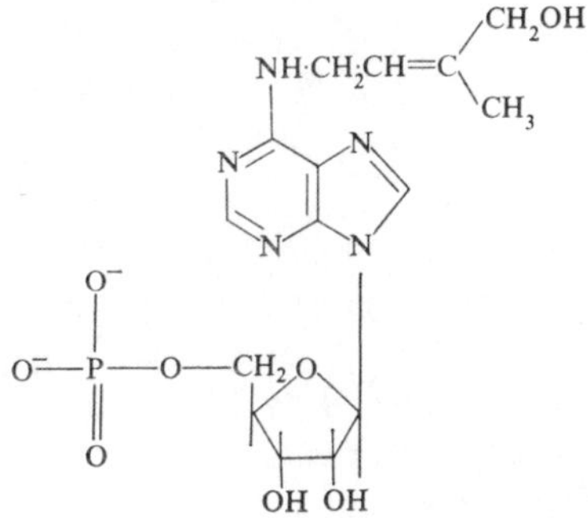

Zeatin ribotide
(9-β-D-ribofuranosylzeatin-5′-phosphate)

Subsequently other cytokinins have been identified. To date, eight cytokinins have been found in corn kernel extracts, but zeatin is the most potent of these. Some, or perhaps all, naturally occurring cytokinins exist in the plant bound to a pentose sugar and sometimes phosphate as well. That is, cytokinins occur as nucleosides or nucleotides. For example, one zeatin derivative found in corn kernels is 9, β-ribofuranosylzeatin-5′-phosphate (a nucleotide).

Natural cytokinins very similar in structure to zeatin and its derivatives have been identified from various sources. Thus, the cytokinin N^6-(Δ^2-isopentenyl) adenosine (a nucleoside) has been obtained from corn kernels, spinach, garden peas, yeast and mammalian tissues. This substance is for convenience called IPA.

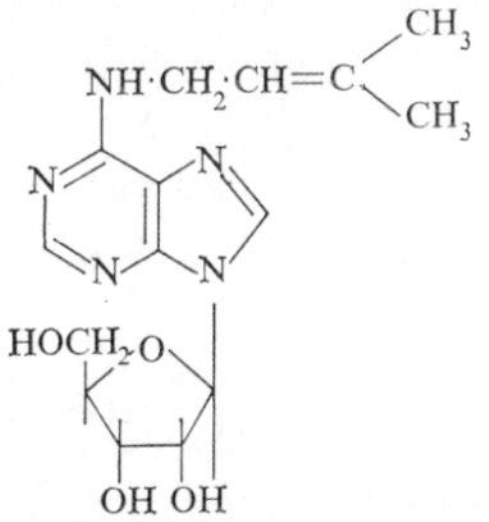

IPA (isopentenyl) adenosine

IPA is now known to occur in transfer ribonucleic acid (tRNA) of plant and animal cells. Therefore, these isoprenoid adenosine compounds exist both in the free form and as their nucleosides and nucleotides. The possible significance of the binding of such compounds in tRNA will be considered later (p. 288).

ETHYLENE

Ethylene may appear a curious substance to consider as a hormone. It is a very simple molecule, contrasting with the more chemically complex gibber-

$$H_2C=CH_2$$

Ethylene

ellins, cytokinins, auxins and abscisic acid. Moreover, ethylene exists as a gas at normal temperatures. Thus, ethylene, if a plant hormone, is a gaseous hormone. It can, and has been, argued that there are theoretical advantages to the plant in having a gaseous, diffusible growth regulator in addition to other hormones which necessarily move through living cells to reach their site of action. At the present time we are not clear as to the extent to which ethylene is involved in normal physiological processes, nor as to the relationship between ethylene effects and those of other known growth hormones. What is certain is that treatment with very low concentrations of exogenous ethylene has many profound effects on physiological and metabolic activities in plants. However, evidence is also accumulating that endogenously synthesized ethylene is involved in the normal control of certain aspects of plant growth, differentiation, and responses to the environment.

It has been known for many years that developing fruits evolve ethylene, and that the time of maximum ethylene production by a ripening fruit coincides with the time of the *respiratory climacteric* (the latter term is applied to the large increase in respiration rate which occurs during the ripening period of many fruits, prior to a fall off in respiration as the fruit enters a senescent decline). It has been found that exposure of fruits to ethylene results in a hastened and enhanced respiratory climacteric with earlier ripening. In fact, ethylene is so effective in stimulating respiration that it does so even in fruits, such as oranges and lemons, which do not naturally experience a respiratory climacteric. The speeding of fruit ripening by ethylene has proved of great commercial value in the citrus industry.

In recent years, however, it has become apparent that we must consider that ethylene has much wider physiological significance than its effects in fruit ripening alone would indicate. In general, it would appear that many effects previously considered to be induced directly by auxin are mediated by an intervening step in which auxin induces the formation of ethylene, following which ethylene induces the actual response. The particular responses which involve ethylene are mainly those which are also known to occur in the presence of relatively *high auxin concentrations*. Although we will return later to considerations of the possible participation of ethylene in physiological processes, it is desirable at this point to catalogue known ethylene effects in plants:

(i) Induction of respiratory climacteric and ripening in fruits.
(ii) Induction of epinasty (see Chapter 7).
(iii) Inhibition of elongation growth in stems and roots of most species,

although ethylene can stimulate elongation in stems, coleoptiles and mesocotyls of certain plants such as *Callitriche* and rice.

(iv) Stimulation of cells to grow isodiametrically rather than longitudinally —thus enhancing radial growth rather than elongation growth in stems and roots.

(v) Stimulation of germination in seeds of some species.

(vi) Induction of root-hair formation.

(vii) Promotion of leaf abscission.

(viii) Induction of flowering in pineapples.

(ix) Induction of flower-fading in pollinated orchids.

(x) Inhibition of basipetal polar and lateral transport of auxin.

(xi) Regulation by a feedback mechanism of the level of endogenous auxin (i.e. high concentrations of auxin stimulate ethylene formation, but the ethylene then, in some way, causes a lowering of the auxin level in the tissue).

It is possible that the concentration of auxin determines the amount and types of protein synthesized; low auxin concentrations induce the formation of those proteins required for active growth, but high auxin concentrations induce, in addition, a protein which catalyses the synthesis of ethylene from its precursor.

Ethylene Biosynthesis

Very little is known of the nature of ethylene biosynthesis. Several compounds, particularly the amino acid methionine, are readily converted by plant tissues to yield ethylene. Light, particularly far-red wavelengths, appears to promote ethylene production, as also does FMN (flavin mononucleotide). Certain evidence available suggests that methionine serves as the natural precursor of ethylene in plants, and that the general pathway of biosynthesis is as follows:

$$CH_3SCH_2 . CH_2CH(NH_2) . COOH \xrightarrow[\text{light}]{\text{FMN}} CH_3SCH_2 . CH_2 . CHO \xrightarrow[\text{light}]{\text{FMN}}$$

(methionine) (methional)

$$CH_2 . CH_2 + (CH_3S)_2 + HCOOH + \text{Other products}$$

(ethylene) (methyl disulphide) (formic acid)

STRUCTURE AND ACTIVITY OF PLANT GROWTH HORMONES

The identification of indole-3-acetic acid (IAA) in 1934 as a naturally occurring plant growth hormone was followed by investigations to determine just what was "special" in the chemical structure of IAA to impart such profound influences on growth and developmental processes. It was hoped that an understanding of this would provide a lead into elucidating the mechanism of action of auxins within the cells. Similar considerations have more recently been applied to the gibberellins and cytokinins following their discovery.

Structure–Activity Relationships of Auxins

The methods of studying this problem were initially largely empirical, in that many compounds were tested in suitable bioassays to find whether or not they possessed any "auxin activity". Some of these substances did in fact produce effects similar to those which IAA itself elicited, even when they were supplied at very low concentrations. Over the years a very large number of such compounds have been found, all of which have been synthesized in laboratories and are, therefore, called *synthetic auxins*. These synthetic auxins do not fall into any one particular class of compound, but despite the diversity of structure shown by the synthetic auxins, very serious efforts have been and still are being devoted to pinpointing exactly what attributes a molecule must possess for it to have activity as an auxin. It is hoped that elucidation of the molecular requirements for auxin activity will help in an understanding of the mechanism by which auxin operates in plant cells.

The first synthetic auxins found were compounds closely related to IAA (in having the indole ring) such as α-(indole-3)-propionic acid, α-(indole-3)-butyric acid and β-(indole-3)-pyruvic acid (Fig. 41) (which is now known also to occur naturally). However, many other synthetic auxins more markedly different in structure from IAA were subsequently discovered, and some of the more active of these, such as 2,4-dichlorophenoxyacetic acid (2,4-D), 2,4,5-tri-chlorophenoxyacetic acid (2,4,5-T) and 4-chloro-2-methylphenoxyacetic acid (MCPA) (Fig. 41), are not indole compounds.

As a result of comparing the structures of all the then available synthetic and natural auxins, in 1938 a list was drawn up of the general structural requirements of a molecule for it to behave as an auxin. Thus, it was said that an active molecule must possess (i) a ring system with at least one double bond present; (ii) a side chain containing a carboxyl group (or group easily converted into a carboxyl-group); (iii) at least one carbon atom between the ring and carboxyl group in the side chain; (iv) a particular spatial relationship between the ring system and the carboxyl group. Later on, it was thought that another requirement of molecules which have activity as auxins is that they must have the ability to form a covalent bond at a position on the ring system *ortho* to the side chain which terminates in a carboxyl group. An examination of the structures of the synthetic auxins shown in Fig. 41, will reveal that they all comply with these general requirements, but there are other compounds now known which possess auxin activity and yet do not fully comply with the above list of structural requirements. For example, certain benzoic acid derivatives are active auxins

(Fig. 43), and yet have no side chain. On the other hand, the activity of certain thiocarbamates indicates that not even the unsaturated ring is essential, although it is necessary that these latter compounds should have a planar structure. Furthermore, instead of the ability to form a covalent bond at the *ortho* position, it is now known that the requirement is that there should be a fractional positive charge at a specific point on the ring (p. 84).

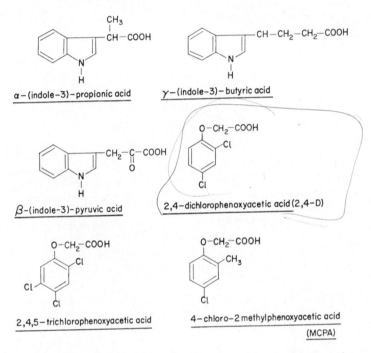

α–(indole–3)–propionic acid

γ–(indole–3)–butyric acid

β–(indole–3)–pyruvic acid

2,4–dichlorophenoxyacetic acid (2,4–D)

2,4,5–trichlorophenoxyacetic acid

4–chloro–2 methylphenoxyacetic acid
(MCPA)

Fig. 41. The chemical configurations of some natural and synthetic auxins.

On the assumption that a carboxyl-terminated side chain, and a "free" *ortho* position on the ring system are essential for activity, it was proposed that the basic reaction of an auxin within the cell involves two parts of the molecule, the carboxyl group of the side chain and an *ortho* position of the ring system. This has led to what is called the *"two-point attachment theory"* for auxin action. The research workers who put forward this theory proposed that there is covalent bond (i.e. chemical bond) formation at these two points between the auxin molecule and some constituent, possibly a protein, of the cell. The principles of the two-point attachment theory and a demonstration of the apparent validity of some of the listed structural requirements for auxin activity, are clearly

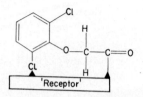

(2,4 — dichlorophenoxyacetic acid)
Active as an auxin — attached
at both receptor active positions

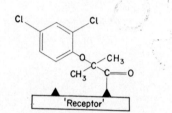

(2,6 — dichlorophenoxyacetic acid)
Inactive — due to both <u>ortho</u> positions
being filled by chlorine atoms.

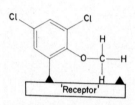

(2,4 — dichloroanisole)
Inactive — due to absence of a
terminal carboxyl group in side chain

(2,4 — dichlorophenoxy <u>iso</u> — butyric acid)
Inactive — spatial configuration prevents
union between receptor active site and
the free <u>ortho</u> position

Fig. 42. The two-point attachment theory of auxin activity, illustrated by a comparison of 2, 4-dichlorophenoxyacetic acid (2, 4-D) with three inactive analogues.

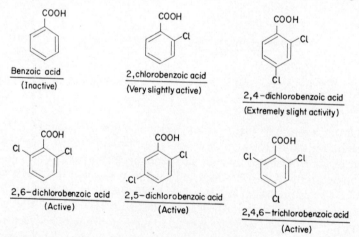

Benzoic acid
(Inactive)

2,chlorobenzoic acid
(Very slightly active)

2,4 — dichlorobenzoic acid
(Extremely slight activity)

2,6 — dichlorobenzoic acid
(Active)

2,5 — dichlorobenzoic acid
(Active)

2,4,6 — trichlorobenzoic acid
(Active)

Fig. 43. Activity or inactivity as auxins of a series of chlorinated benzoic acid derivatives. Note that 2, 6-dichlorobenzoic acid and 2, 4, 6-trichlorobenzoic acid are active and also have a halogen atom at both *ortho* positions.

illustrated by a comparison of some of a series of chlorinated phenoxy compounds (Fig. 42).

It is necessary to stress that neither the original list of requirements for auxin
activity, nor the two-point attachment theory is accepted as valid by all plant
physiologists. Clearly, the activity of a number of synthetic auxins such as benzoic
acid derivatives (Fig. 43) cannot be adequately explained on the basis of the list
of requirements drawn up in 1938 and the two-point attachment theory, and at

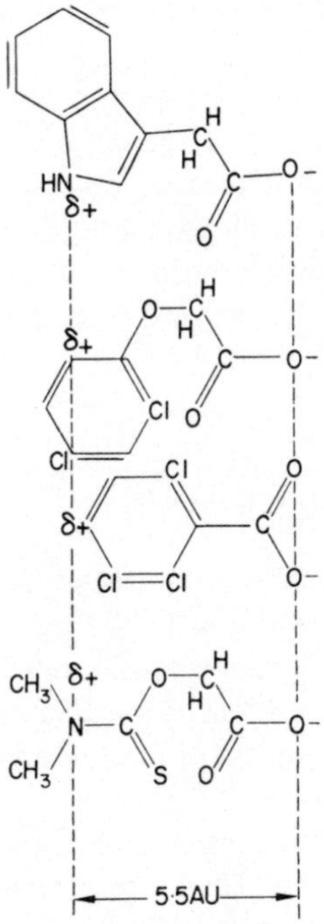

FIG. 44. Some diverse molecules active as auxins. They are, however, similar in the
possession of a strong negative charge ($-$) separated from a weaker positive charge
($\delta+$) by a distance of 5.5 Ångstrom units (5.5ÅU). From top, indole-3-acetic acid,
2, 4-dichlorophenoxyacetic acid, 2,5,6-trichlorobenzoic acid, carboxymethylthio-
carbamate. (From K. V. Thimann, *Ann. Rev. Plant Physiol.* **14**, 1–18, 1963.)

the present time there are several alternative hypotheses, including "three-point"
and "multi-point" attachment theories. The question is still open as to whether
the auxin molecule becomes attached to some receptor in the cell by covalent
(chemical) bond formation, or by some form of physical association. One
suggestion, based on studies of the physical properties of active molecules, is that
van der Waals and electrostatic forces are important in auxin–receptor associa-

tion. Thus, a comparison of a range of auxins revealed that molecules active as auxins contain a strong negative charge (arising from the dissociation of the carboxyl group) which is separated from a weaker positive charge on the ring by a distance of about 5.5 Å (Fig. 44), and it has been suggested that this is the essential structural requirement for auxin activity. This hypothesis would explain the relative activities of many synthetic auxins, and the differences in the activity of closely related compounds are apparently due to the effects of substitution in the ring on the position and size of the positive charge. Neither the nature, nor the location within the cell, of the receptor molecule is yet known.

Thus, the intensive study which has been devoted to the molecular requirements for auxin activity has not yet given any clear indication of the basic mechanism by which auxins produce their effects in growth and differentiation. It has, however, led to results of practical importance in the finding of a number of compounds which have proved of enormous value in agriculture and horticulture, such as selective weed-killers, fruit-setting agents and rooting-hormones (p. 128).

Structure–Activity Relationships of Gibberellins

The relationship between molecular structure and biological activity of gibberellins has been less rigorously studied than it has for auxins. Reasons for this are: (i) the shorter time that has elapsed since the discovery of gibberellins than of auxins, and (ii) it has so far proved impractical to manufacture any "synthetic gibberellins" due to the complex nature of the gibbane carbon skeleton.

As stated earlier in this chapter, well over twenty chemically characterized gibberellins are currently known. All of these have been obtained from natural sources, either from the fungus *Gibberella fujikuroi* or from higher plants. The structures of gibberellins A_1 to A_{13} inclusive are shown in Fig. 45. It can be seen that they are all similar in the possession of the basic gibbane carbon skeleton. The structural differences between them lie principally in the number and distribution of hydroxyl (–OH) groups, and the degree of saturation of the "A" ring. Since no synthetic gibberellins are yet available, present knowledge allows us only to assume that for a molecule to behave as a gibberellin it must have a structure similar to that of known naturally occurring gibberellins. On the other hand, a naturally-occurring diterpenoid in plants called *steviol*, which does not have the gibbane carbon skeleton, has been found to have some slight growth-promoting properties similar to those of gibberellins. However, this is probably due to conversion of steviol by plant enzymes to an active form, rather than to its having hormonal activity itself.

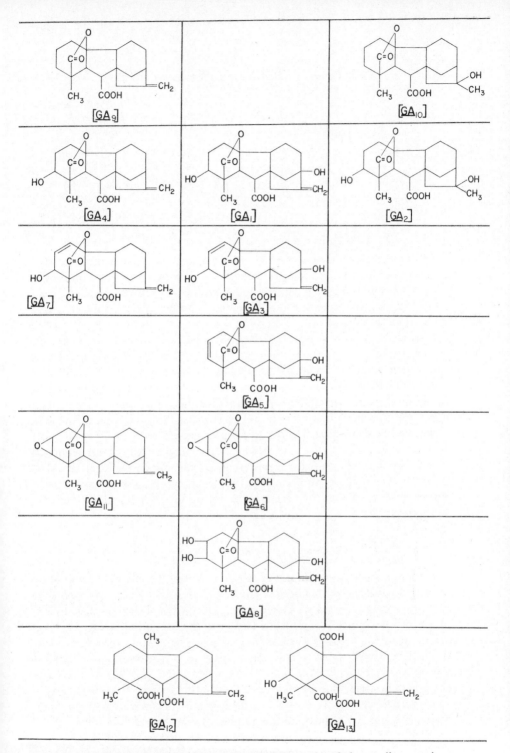

Fig. 45. The chemical structures of the first thirteen identified naturally-occurring gibberellins (GA_1–GA_{13}). The compounds are arranged to show structural relationships. Horizontal groupings indicate similarity in the 'A' ring. Vertical groupings show similarity in substituents on the 7 and 8 positions on the gibbane carbon skeleton. (From L. Paleg. *Ann. Rev. Plant Physiol.* **16**, 291–322, 1965.)

Steviol   Ecdysone

Earlier reports that the insect moulting hormone ecdysone, a steroid, was able to promote growth in a similar manner to gibberellins are now generally regarded as misleading. Subsequent studies with pure preparations of ecdysone have demonstrated that the insect hormone is unable to mimic the effects of gibberellins in plants.

It should be noted that not all of the known gibberellins are equally effective in stimulating growth. In fact, their activity when tested with different species and varieties of plants can be used as a means of distinguishing between different gibberellins. A good example of this is seen in their effect on the growth of dwarf mutants of maize (*Zea mays*). In maize there are a number of mutant genes, the presence of any one of which results in a dwarf habit of growth. Certain gibberellins have been found to promote the growth of some dwarf mutants, while others are effective with other mutants. It has been suggested that the primary effect of the mutant genes in maize is on the levels of endogenous gibberellins, by interfering with different steps in the biochemical pathway leading to a gibberellin necessary for normal growth, though this is not at all certain (p. 102).

Structure–Activity Relationships of Cytokinins

The known natural cytokinins are substituted adenine compounds. A number of synthetic substances have been shown to possess cytokinin activity, and these include not only adenine derivatives, such as kinetin, but also substituted phenylureas which are quite different from adenine in their chemical nature.

Systematic studies of up to a hundred analogues of kinetin have revealed that the adenine moiety is essential for hormonal properties, but that the furfuryl side chain can be replaced by other groups. However, the length of the side chain is critical and the presence of a double bond in the side chain usually increases activity. Where the side chain does not contain a ring system (e.g., zeatin, IPA, see p. 77) the optimum number of side chain carbon atoms is five. Increasing the side chain length to include ten carbon atoms (e.g. 6-decylaminopurine) reduces cytokinin activity to a barely detectable level. One of the most active synthetic cytokinins is 6-benzylaminopurine (BAP), in which the side chain is a benzene ring, and which has been widely used in experimental studies.

NH·CH₂

6-benzylaminopurine (BAP)

Not only 6-substituted adenines, but also a number of 1-substituted adenines have been found to show cytokinin activity. However, it is possible that the 1-substituted compounds may be enzymatically converted to the 6-substituted adenines in plant tissues, and hence the apparent activity of the former compounds may be a result of their chemical change. The minimum requirements for activity in the phenylureas is the –NH–CO–NH– bridge and a planar phenyl ring. At first sight it is difficult to see what structural features these compounds have in common with adenine derivatives, but one common feature is the –N–C–N– linkage, of which adenine derivatives have four and the ureas one. The six-membered pyrimidine ring of adenine may be analogous to the phenyl ring of the ureas and the amino nitrogen of the purine is analogous to a *meta* substituent in phenylurea.

FURTHER READING

General

AUDUS, L. J. *Plant Growth Substances*, 2nd ed., L. Hill Ltd., London, 1959.

LAETSCH, W. M. and R. E. CLELAND (Ed.). *Papers on Plant Growth and Development*, Little, Brown & Co., Boston, 1967.

LEOPOLD, A. C. *Plant Growth and Development*, McGraw-Hill, New York, 1964.

WILKINS, M. B. (Ed.). *Physiology of Plant Growth and Development*, McGraw-Hill, London, 1969.

More Advanced Reading

CROSS, B. E. Biosynthesis of the gibberellins, in *Progress in Phytochemistry* (Ed. L. Reinhold and Y. Linschitz), Interscience Publishers, London and New York, 1968.

GALSTON, A. W. and W. S. HILLMAN, The degradation of auxin. *Encyc. Plant Physiol.* **14,** 647.

GORDON, S. A. The biogenesis of auxin. *Encyc. Plant Physiol.* **14,** 620.

HESLOP-HARRISON, J. Plant growth substances, in *Vistas in Botany*, Vol. 3 (Ed. W. B. Turrill), Pergamon Press, Oxford, 1963.

MILLER, C. O. Kinetin and related compounds in plant growth. *Ann. Rev. Plant Physiol.* **12,** 395, 1961.

PALEG, L. G. Physiological effects of gibberellins. *Ann. Rev. Plant Physiol.* **16,** 291, 1965.

SHANTZ, E. M. The chemistry of naturally-occurring growth-regulating substances. *Ann. Rev. Plant Physiol.* **17,** 409, 1966.

THIMANN, K. V. Plant growth substances; past, present and future. *Ann. Rev. Plant Physiol.* **14,** 1, 1963.

Various articles in:

(a) *Regulateurs Naturels de la Croissance Végétale* (Ed. J. P. Nitsch), Centre Nat. Recherche Scient., Paris, 1964.

(b) *Biochemistry and Physiology of Plant Growth Substances* (Ed. F. Wightman and G. Setterfield), The Runge Press, Ottawa, 1969.

CHAPTER 5

The Role of Hormones in Shoot and Root Growth

THE distribution and pattern of growth and differentiation in the whole plant was considered in the first three chapters of this book. Clearly, a plant's rate of growth and the pattern of its development depends upon both its genetic constitution, or genotype, and on environmental factors. We may visualize control of growth and differentiation as either *internal*, involving the "programmed" pattern of differentiation described in earlier chapters, or *external*, when it involves environmental factors, such as daylength and temperature (Chapters 9 & 10). Internal control may be either (1) intracellular, or (2) intercellular. Growth hormones appear to have a particular role in control at the intercellular level, serving to correlate growth in spatially distinct regions of the plant body. Nevertheless, once a growth hormone enters a cell it influences a number of biochemical activities which constitute part of the intracellular control mechanism (see Chapter 13). Also, many environmental effects on plants are apparently mediated through changes in internal hormone levels and distribution. Both genetic and environmental control of growth and differentiation are, therefore, achieved by means which include the participation of growth hormones.

We may reasonably assume that the distribution and concentration of auxins, gibberellins and cytokinins, together with other types of hormones such as abscisic acid and ethylene, are vital factors in the overall control of growth and differentiation in the whole plant. Nevertheless, the pattern of distribution of growth hormones in the plant with regard to space, time and concentration, is itself controlled by interactions between the environment and the genetic make-up of the plant (genome). That is, growth hormones serve merely as agents, albeit very influential agents, in the overall integration and co-ordination of growth and differentiation.

The influence of growth hormones extends to all phases of development of the whole plant, including germination and dormancy, flowering, senescence and growth movements. In this chapter we consider what is known of the role of growth hormones in various aspects of vegetative growth and differentiation. A similar consideration of growth hormones and fruit development is made in Chapter 6, and the possible significance of the same substances in the processes of flower initiation is considered in Chapter 10.

Although it will be convenient to consider separately the effects of hormones in various phases of growth and differentiation of both shoots and roots, in the living plant, these various effects are not kept neatly in separate compartments but they represent different aspects of a continuous and co-ordinated series of changes, in which hormones play essential roles in the overall co-ordination of growth in the plant as a whole. Two very important factors in the co-ordination of growth in the whole plant are (1) the localization of the sites of hormone biosynthesis, and (2) the pattern of hormone movement and distribution from these sites.

The young tissues, especially the developing leaves, in the shoot apical region are active sites of auxin biosynthesis but it appears that some auxin biosynthesis also occurs in older leaves. Developing ovules also constitute another site of biosynthesis. There is still doubt as to whether auxin is synthesized in the roots. It appears that, in general, auxin is transported from the centres of production in the shoot in a predominantly basipetal (downwards) direction (p. 96), but, the mechanism of this "polar" transport process is still not understood.

Gibberellins appear to be synthesized in growing leaves, fruits and roots, and to be freely transported in all directions in the plant, probably in the phloem and xylem. The evidence available at present suggests that biosynthesis of cyto-kinins may be restricted to the roots (p. 265); developing seeds are rich in this type of hormone, but whether they are synthesized there or are accumulated there from other parts of the plant, is not known. We know little about the normal path of transport of cytokinins, but they appear to be present both in the xylem sap and in the phloem.

As a result of the high rate of auxin biosynthesis in the shoot apical region and its basipetal transport from there, a *gradient* of auxin concentration is present in the plant; probably an important additional factor in the establishment of these gradients is the fact that a proportion of the transported auxin becomes im-mobilized or destroyed (p. 69) in the tissues through which it passes. Gradients of gibberellin concentration also appear to be present in the stem, with the highest concentrations in the apical region. We shall see that there is a correla-tion between growth rates in different regions of the stem and the gradients in auxin and gibberellin concentrations. Thus, growth in the stem internode is evidently affected, and may be controlled, by hormones entering from the shoot apical region. Moreover, we have seen (p. 59) that young, developing leaves also have an important effect on the differentiation of vascular tissue in the stem and it is now clear that this effect is mediated by auxins and gibberellins arising in the young leaves (p. 108). We shall later meet other examples of such *growth correlations*, in which hormones appear to play important roles (Chapter 6).

We are much more clearly aware of the importance of hormones in the control of growth and differentiation in existing organs, than we are of any part they may play in the formation of organs in the intact plant. Nevertheless, there is at least the possibility that auxins, gibberellins, cytokinins, abscisic acid and ethylene do help to determine the sites of organ initiation, and trigger off the necessary reactions for this process, though there is little or no direct evidence that this is so. What is suggestive, but no more, is that known growth hormones can promote the formation of shoot buds and roots in callus cultures (Chapters 4 and 8).

MODE OF ACTION OF AUXIN IN CELL EXTENSION

As we saw in Chapter 1, all plant cells, with the exception of those which remain permanently meristematic, pass through two phases in their cycle of growth; these are division and enlargement resulting from vacuolation. The coleoptile of an oat seedling illustrates this pattern in a clear manner, since all cell division ceases when it is about 10 mm in length and all subsequent growth is entirely due to the enlargement of existing cells. Hence, when we study the effects of auxin on the growth of coleoptile sections (p. 66), we are essentially dealing with hormonal effects on cell extension. During cell enlargement, due to vacuolation, irreversible plastic stretching of the cell wall takes place. It is, therefore, tempting to consider that cell vacuolation is a consequence of a softening of the cell wall, for this would inevitably lead to an influx of water into the protoplast for the reasons given in Chapter 1 (p. 7). Many experiments have revealed that auxin increases the plasticity of the cell walls. This can be shown by increased plastic deformation of plant organs treated with auxin following the application of a mechanical force (Fig. 46). The possible physiological significance of auxin effects on cell wall plasticity is increased by observations that there is a positive correlation between the effects of different auxin concentrations on promotion of elongation growth, and on cell wall plasticity (Fig. 47).

During cell enlargement, the cell wall not only stretches, but it also increases in thickness by the deposition of new cell wall material (p. 9). This cell wall growth is stimulated by auxin, and can occur even when cell enlargement is completely suppressed by various means (e.g. by surrounding the tissue with a hypertonic solution of mannitol).

How do auxins influence the physical properties of cell walls? As we saw in Chapter 1, the walls of young growing cells consist of interwoven chains of

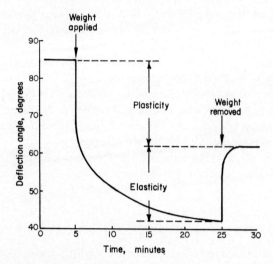

FIG. 46. Plastic deformation of an oat coleoptile following application of mechanical force. The coleoptile was held horizontally and a weight hung on one end for twenty minutes. The weight was then removed, and the difference between original and final positions of the coleoptile (in degrees) was taken as a measure of the plasticity. (From T. Tagawa and J. Bonner, *Plant Physiol.* **32,** 207–12, 1957.)

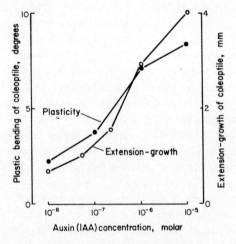

FIG. 47. Positive correlation between effects of auxin (IAA) on cell wall plasticity (measured by plastic bending as in Fig. 46) and on elongation in the oat coleoptile. (From J. Bonner, *Z. Schweiz. Forstv.* **30,** 141–59, 1960.)

cellulose microfibrils embedded in an amorphous matrix of non-cellulosic polysaccharides. In general, the cell wall can be regarded as being stabilized by the properties of the matrix. Consequently, it has been suggested that changes in cell wall plasticity are a consequence of an alteration in the binding properties of the polysaccharides of the matrix.

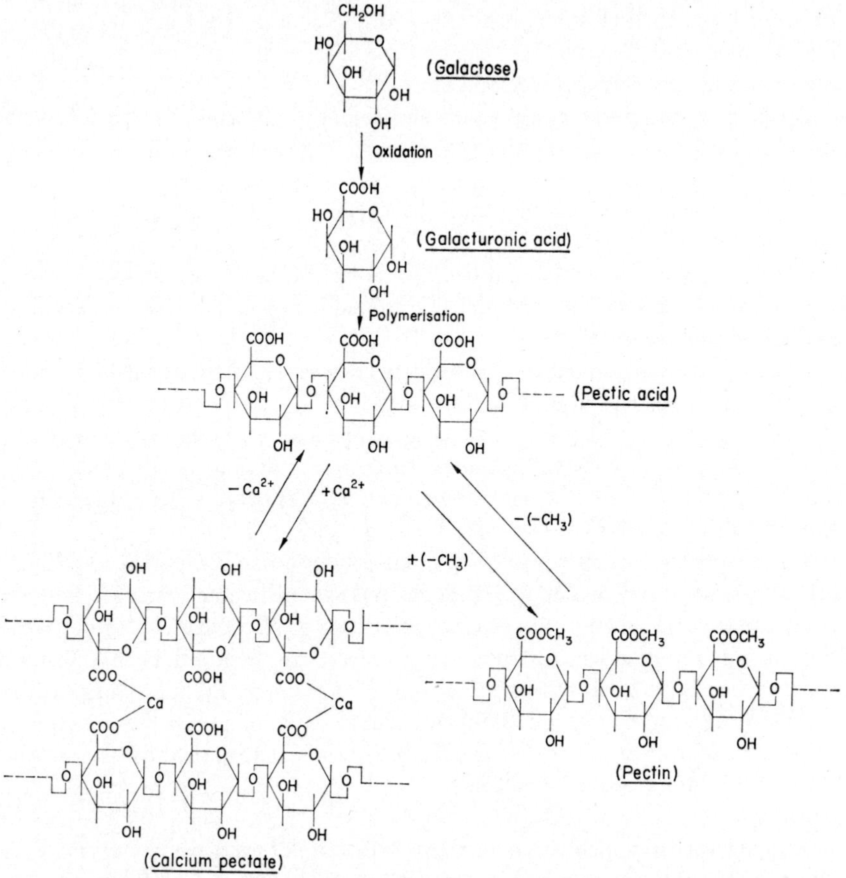

FIG. 48. Formation and interconversion of some pectic substances in plant cell walls.

The non-cellulosic polysaccharides of the cell wall are very heterogeneous and complex, and their structure is not yet fully elucidated. In fact, there appears to be a continuous "spectrum" of polysaccharides from those containing a high percentage of uronic acids, the "pectic" substances, to those with a low content of uronic acids, the "hemicelluloses".

Until recently, it was thought that pectic acid consisted solely of 1–4–linked galacturonic acid residues. Galacturonic acid is derived from the hexose sugar, galactose, by oxidation of carbon–6 from a carbinoyl group ($-CH_2OH$) to a carboxyl group ($-COOH$) (Fig. 48). It now appears that other sugars are attached to the polygalacturonide "backbone" and may even form part of the main chain, so that "pectic acid" is of variable composition. Many carboxyl groups of the galacturonic acid may be esterified by the transfer of methyl ($-CH_3$) groups from a methyl donor such as the amino acid, methionine. Thus, pectic acid is methylated to yield its methyl ester, "pectin". Pectins may be re-converted to pectic acid by hydrolysis.

Apart from the pectic components and cellulose, there are, in the cell wall, polysaccharides based upon other sugars, including pentoses, such as xylose and arabinose, and hexoses, such as glucose, galactose and mannose, which may form linear chains constituting the "backbone", to which other sugars are attached. Indeed, usually none of these polysaccharides consists exclusively of one type of sugar residue, but contains a variety of others. Such polysaccharides are commonly included in the rather ill-defined group of substances referred to as "hemicelluloses".

Much of the pectic acid in cell walls is apparently in the form of calcium pectate. Thus, the addition of calcium ions (Ca^{2+}) has the effect of increasing the rigidity of cell walls, for cross-linkages are formed between adjacent chains of calcium pectate, with the bivalent calcium ions forming the "bridges" (Fig. 48). In fact, calcium inhibits cell enlargement and this may be due to the forma-tion of calcium pectate in the wall. Consequently, it has been suggested that auxin in some way removes calcium from pectic substances, perhaps by bind-ing with Ca^{2+}, and in this way induces cell wall softening. Support for this proposition came when it was observed by some workers that chelating agents such as EDTA (ethylene diamine tetra-acetic acid) induced both cell wall softening and cell enlargement in coleoptiles. However, the stimulating effect of EDTA on growth was found not to occur in experiments with other tissues. Other evidence against a chelating effect of auxin was furnished by the finding that auxin did not remove radioactive Ca^{2+} previously incorporated into cell walls.

Another way in which auxin may affect the state of pectic substances is by an effect on the formation of pectin from pectic acid. A number of experiments demonstrated that auxins increase the incorporation of methyl groups from methyl donors such as methionine or formaldehyde, into cell wall pectin. Clearly, calcium is not able to form cross-links between pectin molecules in the way that it does with pectic acid, and a high level of pectin in the cell wall would be reflected in a high degree of cell wall plasticity. Attractive though the pro-position might be, it appears that stimulation of cell enlargement by auxin does not result solely from stimulation of pectin formation, for it has been found that auxin is still able to induce cell elongation even under conditions where methylation of pectic substances is completely prevented (e.g. by the addition of ethionine).

Apart from changes in the pectic components, it is becoming increasingly clear that plant cells contain enzymes which are able to bring about hydrolysis of cell wall components, including both pectic and hemicellular fractions. Thus, it is possible that increased plasticity of the cell wall during growth may be due to the activity of such hydrolytic enzymes, which will break down the glycosidic linkages within the polysaccharide matrix. Evidence in support of this hypothesis is provided by the finding that the growth of *Avena* coleoptile segments is increased by treating them with the enzyme, β-1,3-glucanase, which hydrolyses β-1,3-glucose links in the cell wall of coleoptiles. It was also shown that the enzyme increased the plastic extensibility of the walls.

In further experiments with *Avena* coleoptiles, an enzyme system was found which was capable of hydrolysing polysaccharides and which was bound to the cell walls. IAA was shown to increase the activity of this enzyme system. Thus, it is possible that IAA may stimulate the synthesis or activity of enzymes capable of hydrolysing polysaccharides and hence bring about increased plasticity of the wall in this way. There is evidence that the enzyme cellulase may be involved in lateral expansion of cells during vacuolation, by hydrolysing and cleaving the cellulose microfibrils, which, as we have seen (p. 11) tend to be horizontally oriented in the early stages of growth and hence will resist lateral expansion.

Recent work has shown that protein constitutes an important component of the cell wall. Some of the cell wall protein is enzymic and includes not only the hydrolytic enzyme ("hemicellulase") already referred to, but other enzymes, including invertase, pectin methylesterase and ascorbic acid oxidase. However, in addition to these enzymes, which appear to be bound to the polysaccharide components, most cell walls also contain a very characteristic protein or proteins rich in the amino acid, hydroxyproline. This protein is apparently not an enzyme, since it has not been found to show any enzymic activity. Its function is not fully known but it seems very likely that it may play an important structural role in the wall; if this is so, then it is very likely to be involved in changes in wall plasticity and cell extension.

It has, in the past, been suggested that the plasticizing effect of auxin on the cell wall is the primary point of auxin action in the cell, but this theory is certainly too simple. In fact, although cell enlargement is often the most characteristic result of auxin action, it is by no means the only one, nor even the first to appear. Thus, auxin induces a rapid increase in the rate of respiration, of protoplasmic streaming, and of nucleic acid synthesis (see Chapter 13). Also, a number of responses to auxin do not immediately involve cell vacuolation (e.g. cambial division, root initiation, and correlative inhibition of axillary buds). It would seem, therefore, that the effect of auxin on cell wall plasticity is a secondary one.

POLAR TRANSPORT OF AUXINS

Auxin translocation in shoot and root tissues is polarized (i.e. auxin moves more rapidly in one direction than in the opposite direction). The polar transport of auxin is undoubtedly of great importance in the co-ordination of growth and development in different parts of the plant.

In the shoot tissues studied (coleoptiles, stems, hypocotyls and petioles), auxin moves more rapidly *basipetally* (i.e. from apex to base) than *acropetally* (from base to apex). As we shall see later, auxin transport is also polar in roots, but here movement is preferentially acropetal.

Auxin Transport in Shoot Tissues

Polar basipetal auxin transport occurs in all organs of the vegetative shoot. The majority of experiments which have shown this have been conducted with short excised segments (usually 5–10 mm long) of coleoptiles, stems, petioles, etc. In principle, the technique is to apply an auxin to one end and to follow its movement along the segment. Various methods have been adopted to determine how much auxin has been transported, and how far, in such segments, but usually a "donor–receiver" agar block system is employed. In this, an agar block containing auxin (the "donor block") is placed against one cut end of a segment of tissue, and another agar block (the "receiver block") against the opposite end. Auxin molecules enter the segment from the donor block, are transported through the segment and eventually emerge into the receiver block. Once auxin starts to enter the receiver block, its concentration there rises linearly with time under carefully controlled experimental conditions. The intercept on the time axis of the straight line of increase in auxin content of the receiver block provides an estimate of the average time taken for auxin molecules to pass from one end of the segment to the other (Fig. 49A). Since the length of the segment is known, auxin movement can be expressed in terms of velocity (distance moved in unit time).

Using the donor–receiver block method, Went, in 1928, found that auxin moved only basipetally in *Avena* coleoptile segments. Irrespective of the orientation of a segment with respect to gravity, auxin appeared only in a receiver block placed against the morphological basal end with a donor block positioned at the morphological apical end. The auxin used in Went's experiments was unknown, but it was collected from *Avena* coleoptile tips and was probably IAA. Other investigators repeated Went's experiment, and confirmed the existence of polar basipetal IAA transport in coleoptiles, stems, hypocotyls and petioles. Earlier workers, like Went, measured the quantity of IAA in receiver blocks by bioassay. More recently, the availability of radioactive auxins has allowed more precise experimentation, and this has revealed that (a) auxin transport in aerial organs is not exclusively polar, for some acropetal as well as basipetal movement takes place (Fig. 49B), and (b) in addition to IAA, certain synthetic auxins, such as 2,4-dichlorophenoxyacetic acid (2,4-D),

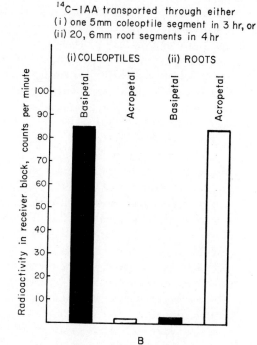

Fig. 49A. Estimation of the velocity of basipetal polar transport of indole-3-acetic acid (IAA) in bean petiole segments. Radioactive IAA (^{14}C-IAA) at a concentration of 50 μM in agar-gel (the 'donor block') was supplied to the apical end of each segment, and a blank agar 'receiver block' was placed against the basal end. Radioactivity appearing in the receiver block was determined at hourly intervals. The graph line showing radioactivity present in the receiver block intercepts the time axis at 0.8 hour. The petiole segments were 5.44 mm long, which means that IAA was transported basipetally at velocity of 6.8 mm/hour. (From C. C. McCready and W. P. Jacobs, *New Phytol.* **62,** 19–34, 1963.)

B. Polar transport of indole-3-acetic acid (IAA) in coleoptile segments (i), and in root segments (ii) of *Zea mays* c.v. "Giant White Horsetooth". Donor agar-blocks containing radioactive IAA (^{14}C-IAA) were placed on either the apical end (basipetal transport) or basal end (acropetal transport) of coleoptile or root segments. The amount of radioactivity (counts per minute) which appeared in blank agar receiver blocks at the opposite ends of the segments was determined. In coleoptiles polar auxin transport is basipetal, but in roots the direction of movement is acropetal. (Coleoptile data from M. B. Wilkins and P. Whyte, *Planta (Berl.),* **82,** 307–16, 1968; root data from M. B. Wilkins and T. K. Scott, *Nature,* **219,** 1388–9, 1968.)

indole-3-butyric acid and naphthalene acetic acid (NAA) are transported in a polar manner.

The velocity of basipetal polar auxin transport has been measured in various organs by a number of workers. Values obtained for IAA polar transport all lie between 5 and 15 mm per hour. Synthetic auxins, although transported in a polar manner, apparently move more slowly. For example, 2,4-D moved basipetally in *Phaseolus vulgaris* petiole segments at a velocity of only 1 mm per hour, whereas the equivalent figure for IAA was 6 mm per hour.

The velocity of acropetal auxin transport has not been rigorously determined for aerial organs, but it is normally very much lower than that of basipetal transport. However, the differential between the velocities of basipetal and acropetal auxin transport is influenced by a number of factors. Thus, polarity of auxin movement declines with increasing age of the transporting tissue. It is not yet clear whether this is due to a decrease in basipetal transport or an increase in acropetal transport, or both. There is, nevertheless, no doubt that maturation processes in a tissue are associated with a gradual reduction in the polarity of auxin transport. Because of this, it has been suggested that polar transport of auxin occurs only in association with cell elongation. However, careful experiments by McCready and Jacobs in 1967 showed that, in bean petiole segments, basipetal polar transport of 2,4-D was less when the segments were elongating rapidly in the presence of gibberellic acid (GA$_3$) than when their growth was inhibited by mannitol. On the other hand, earlier experiments demonstrated that GA$_3$ stimulated basipetal IAA transport in stem tissues. Thus, we know neither the significance nor the basis of reduced polarity of auxin transport in mature tissues.

Polar auxin transport is an active process (i.e. it depends upon metabolically derived energy). Both aerobic and anaerobic respiration appear to be involved, for although a reduction of oxygen tension drastically reduces basipetal transport of IAA in coleoptile segments, the process is not completely prevented even in an atmosphere of pure nitrogen. However, in the presence of metabolic inhibitors, such as sodium fluoride or iodoacetate, polar transport of auxin is completely abolished, which demonstrates that the residual polar transport which occurs under anaerobic conditions is due to energy derived from anaerobic respiration. Acropetal transport of auxin in shoot tissues is unaffected by anaerobic conditions or by metabolic inhibitors, and it is clear that this slow non-polar movement of auxin represents physical diffusion of auxin molecules within the continuous aqueous phase of plant tissues.

One substance, triiodobenzoic acid (TIBA), is an extremely effective inhibitor of polar auxin transport and is often used in experiments for this purpose. Both gibberellic acid and kinetin have been reported to increase the velocity of auxin transport in stems, but the general significance of these observations has not been established.

The effects of temperature on polar auxin transport have not been carefully studied, but it appears that most rapid transport occurs at temperatures known to be favourable for the activities of most enzymes (i.e. 20–30°C).

Gravity appears to have some influence on basipetal polar transport of auxins, for several workers have found that when a normally erect organ is placed horizontally, or inverted, then the velocity of basipetal auxin transport is reduced. This phenomenon may be involved in geotropic responses of plant organs, although much more work needs to be done to evaluate this possibility.

Despite detailed studies of polar auxin transport, extending over 40 years, we still do not know the pathway of auxin movement. The velocity of polar auxin transport (0·5–1·5 cm per hour) is very much less than that of solute movement in the phloem (10–100 cm per hour), and the direction of solute flow in the phloem in the upper stem is acropetal rather than basipetal. For this and other reasons, it does not seem likely that auxin normally travels in the phloem. The other vascular tissue, the xylem, clearly does not usually serve as a transporting channel for auxin, for here again the flow is upwards, and dead xylem elements would be unable to provide the energy required for polar auxin transport. Early work indicated that *all* cells in coleoptile segments are capable of transporting auxin basipetally at 1 cm per hour. We cannot be certain that this is true for stems, for Jacobs found that *Coleus* stem pith-segments failed to transport IAA at all unless a strand of vascular tissue was present. More work needs to be done, therefore, to establish the pathway of polar auxin transport.

Auxin Transport in Roots

Until relatively recently, very few direct studies were made of auxin movement in roots, and perhaps because of this, a great deal of confusion existed over this matter for many years. However, experiments by several groups of workers since 1964 with root segments, using radioactive IAA and the donor–receiver agar block method, have amply confirmed that roots of a range of species show polar transport of auxin, and that the direction of movement is *acropetal* (Fig. 49B). This is the reverse of the situation in shoot tissues, and the full physiological significance of this with respect to the control of root growth and geotropic behaviour must now be evaluated. The velocity of polar acropetal auxin transport in root segments of *Convolvulus arvensis* has been found to be approximately 1 cm per hour, which is similar to the velocity of basipetal polar auxin transport in shoot tissues.

GROWTH HORMONES AND STEM GROWTH

We shall now examine the evidence for the role of hormones, especially auxins and gibberellins, in the control of stem growth.

Auxin and Internode Elongation

If the apical end of a coleoptile is cut off, extension growth of the young cells slows down or stops, which means that elongation of the coleoptile as a whole

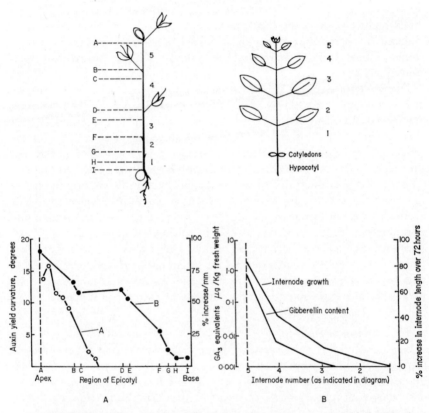

FIG. 50A. The correlation between growth rate and auxin content along the length of a pea seedling stem (epicotyl). *Above*: diagram of a green nine-day-old "Alaska" pea seedling. The figures indicate the number of each internode, and the letters delimit regions assayed for auxin content. *Below*: distribution of diffusible auxin and extension growth in the epicotyl. Curve A, relative growth rates. Curve B, auxin yields. (From T. K. Scott and W. R. Briggs, *Amer. J. Bot.* **47,** 492–9, 1960.)

B. The relationship between elongation rate and gibberellin content along the length of the stem of a young sunflower (*Helianthus annuus*) plant. (From R. L. Jones and I. D. J. Phillips, *Plant Physiol.* **41,** 1381, 1966.)

also ceases. However, if one excises the tip of an elongating coleoptile, and immediately applies an auxin such as IAA to the cut surface of the stump in some suitable medium (e.g. agar gel or lanolin), then extension growth of the coleoptile continues for a while unimpaired. This would seem to indicate that the tip of the coleoptile normally supplies auxin to the newly formed cells, and that the auxin is necessary for the enlargement of those cells.

In a plant stem, it is the region of young tissue behind the apical meristem which shows the greatest extension, the tissues further away from the apex having effectively completed their extension growth. Measurements of growth and auxin along the length of pea seedling (*Pisum sativum*) stem have shown that there is a positive correlation between growth rates in different regions and the quantities of diffusible auxin (see p. 66) obtainable from them (Fig. 50). The correlation between extractable auxin (see p. 66) and growth rate in the pea stem has been found to be less close. On the other hand, extractable auxin of the oat coleoptile has been found to correlate well with growth rate in that organ. These observations are consistent with the view that the auxin concentration in a tissue may regulate its growth rate.

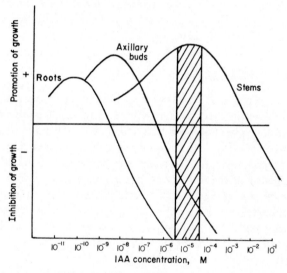

FIG. 51. Growth responses of roots, stems and buds to different auxin concentrations (log scale). Shaded area indicates range of optimal auxin concentrations for stem growth.

Further evidence that auxin is concerned in the control of stem growth is afforded by experiments in which isolated segments of internodes are used. Such segments have been deprived of their supply of auxin from the apical region of the shoot, and consequently the effects of known concentrations of auxin can be measured. Thus, if the internode segments are floated in an appropriate solution of auxin, then they will grow more than when placed in water alone. Further, the amount of extension growth that occurs is proportional to the logarithm of the concentration of auxin in solution (Fig. 51). It can be seen that stem segments show increased growth with increasing concentrations of IAA up to 10 mg/litre ($5 \cdot 7 \times 10^{-5}$ M). On raising the concentration of applied IAA above 10 mg/l less growth takes place, until at concentrations of 100 mg/l and above, there is *inhibition* of extension growth of dark-grown internode sections by applied IAA (Fig. 51). Consequently, we can say that for etiolated stem tissues an IAA concentration of approximately 10 mg/l is *optimal* for cell enlargement. Concentrations greater than this are, therefore, *supra-optimal* and lower concentrations are *sub-optimal*. In some species, such as pea, much higher auxin concentrations are required to produce supra-optimal inhibition in light-grown stem segments than with etiolated stem segments of the same species.

Since application of auxin to an intact shoot (e.g. as a spray) rarely stimulates stem elongation, it may be assumed that the stem tip normally supplies sufficient auxin to the elongating internodes to maintain an optimal auxin concentration in those tissues. The possible reasons for inhibition of growth by auxins will be discussed below (p. 128).

Gibberellins and Internode Elongation

There are several pieces of evidence which indicate that gibberellins, as well as auxins, are involved in extension growth of plant tissues. The most striking and characteristic response of a plant treated with a gibberellin such as gibberellic acid is that stem elongation is stimulated, so that the treated plant becomes taller than it would normally. This response of the stem is usually due to an increased elongation of the internodes and there is generally no increase in the number of internodes formed. The increased internode length is a consequence of increased cell extension and cell division. Thus, gibberellin treatment of intact plants can cause enhanced elongation of existing internodal cells, and also increase the number of cells present in each internode, principally as a result of an increase in mitoses in the sub-apical region of the stem (Fig. 52).

The magnitude of the stem-elongation response to gibberellin varies from species to species and from variety to variety within a species. As mentioned in Chapter 4 (p. 71), the response is greatest in genetically dwarf plants—so much so, that following gibberellin treatment dwarf varieties often grow to the same height as that of related tall varieties (Fig. 39). Tall varieties of the same species respond only slightly or not at all. The responsiveness of different varieties to applied gibberellin may perhaps be related to the amount of endogenous gibberellin present in the tissues. However, in some species, such as *Pisum sativum*, contradictory results have been obtained by different research groups in that some have found lower levels of gibberellins in dwarf than in tall pea varieties, whereas others have been unable to detect any quantitative difference. Consequently, one cannot yet conclude that dwarfness of a plant is always due to impaired gibberellin synthesis, and further research is being conducted to resolve this question.

It is possible to extract gibberellins from various organs, and to compare the amounts present in them; also, it is possible to collect gibberellins from organs by the agar diffusion technique previously used in studies of auxins. Such studies have revealed that the apical bud of the shoot, and young leaves, synthesize and export gibberellins to the stem. Also, a good correlation has been obtained in sunflower (*Helianthus annuus*) between the growth rates of internodes of different ages and the gibberellin contents of the same internodes (Fig. 50B). Thus, as

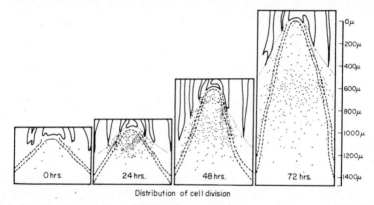

Distribution of cell division

Fig. 52. Stimulatory effect of gibberellic acid (GA₃) upon sub-apical meristematic activity in *Samolus parviflorus*, a rosette plant. Each dot represents a sub-apical cell undergoing mitosis seen in median longitudinal sections of the stem apical region. Twenty-five micrograms of GA₃ were applied 0, 24, 48 and 72 hours Previously. (Reprinted from R. M. Sachs, C. F. Bretz and A. Lang, *Amer. J. Bot.* **46**, 376–84, 1959.)

with auxins, endogenous gibberellins are present in highest concentration in those regions of the stem which are undergoing most rapid extension growth, providing strong circumstantial evidence that gibberellins are concerned in the normal control of stem extension growth. Cytokinins may be involved in the control of cell division rates at the stem apex, but we have no evidence that internode extension is in any way under a direct influence of this class of hormone.

Auxin–Gibberellin Interactions in Stem Elongation

Following the discovery of gibberellins and the realization that these hormones occur naturally in higher plants (see Chapter 4), many plant physiologists were led to study the interactions of gibberellins and auxins in stem and coleoptile extension growth.

Spraying an intact plant with giberellin was found to enhance internode extension growth, whereas it was already known that similar treatment with an auxin rarely induced greater internode elongation. Conversely, if young internodes or coleoptile segments were excised and floated in solution, then it was noted that the opposite usually occurred; auxins stimulated internode or coleoptile segment elongation but gibberellins had little or no effect. However, when gibberellin and auxin were present in solution together, then the elongation of excised segments was much greater than when auxin alone was supplied. In other words, for the characteristic effects of gibberellins on stem elongation to appear, auxin also has to be present. The growth-promoting action of gibberellins when applied to intact plants would, therefore, be a consequence of an interaction between supplied exogenous gibberellin and the natural endogenous auxins present. Because of such observations, it was proposed that gibberellins exert their physiological effects through some "auxin-mediated" mechanism. Thus, there is good evidence that application of gibberellins leads to increased endogenous auxin levels, by effects on either the biosynthesis or the destruction of auxin.

However, it is now clear that gibberellins are a class of growth hormone in their own right. If gibberellin effects were due solely to an influence on auxin activity then it would be expected that these effects would always be the same as those produced by auxins. Whilst gibberellins often are able to duplicate the known effects of auxins (e.g. induction of parthenocarpic fruits, promotion of cambial activity), there are many other examples of gibberellins having physiological effects not possessed by auxins (e.g. promotion of stem elongation in intact plants, breaking dormancy of buds or seeds, stimulation of mesophyll growth), or sometimes instances where gibberellins have the opposite effects of auxins (e.g. auxins promote but gibberellins inhibit root initiation in stem cuttings).

Nevertheless, it is apparent that interactions between auxins and gibberellins occur in many physiological responses apart from extension growth. For example, auxin and gibberellin together are often more effective in inducing the development of parthenocarpic fruits and in stimulating cambial activity (p. 110) than is either type of hormone on its own.

Ethylene and Internode Elongation

Impressive evidence has accumulated in recent years which suggests that ethylene may play some part in the control of stem growth. With most species, exposure of stems or isolated internodes to ethylene reduces cell elongation but enhances isodiametric cell expansion. Such ethylene-treated internodes are shorter and thicker than untreated internodes. These effects of ethylene are similar to those which can be induced by high (supra-optimal) concentrations of auxins. It is possible that auxins are not themselves inhibitors of stem elongation, but rather that at high concentrations they stimulate the synthesis of ethylene in plant tissues, which in turn, suppresses cell elongation.

Nevertheless, this does not necessarily mean that stem elongation is *normally* controlled by ethylene. In the case of etiolated pea seedlings, however, it has been shown that the inhibition of growth in the plumular region is due to the presence of ethylene emanating from a region of the etiolated epicotyl just below the plumular hook itself. This indicates that ethylene, like other growth hormones, may serve as a correlation factor in the control of stem growth. Clearly, much more work is required before we can evaluate the physiological significance of the effects which ethylene has in stem growth, particularly in view of recent reports that ethylene enhances stem, coleoptile and mesocotyl elongation in species such as *Callitriche* and rice.

HORMONES AND ROOT GROWTH

It is surprisingly difficult to give, with any degree of certainty, a clear indication of parts played by growth hormones in root elongation. Earlier workers noted that excision of the root apex does not usually prevent elongation of the newly formed cells in the root, in contrast to the inhibiting effect of decapitating a shoot or coleoptile. In fact, it has on occasion been reported that excision of the root tip can result in temporarily *stimulated* elongation of the young root cells. It might, therefore, be thought that the root tip is not a source of growth hormone for root elongation. However, there is evidence that the root tip does in fact synthesize auxin and that this is involved in the regulation of root extension growth. This has been demonstrated in several ways. For example, segments of the young sub-apical region do show growth responses to exogenous auxin basically similar to those of stem or coleoptile segments (Fig. 51), except that the optimal concentration of IAA for root elongation is very much lower (by at least 100,000 times) than for stems or coleoptiles. Also, the extent of promotion of root elongation by added auxin is less than in stems or coleoptiles. In general, it appears that the root tip supplies auxin in amounts that are certainly not limiting root elongation, but may under some conditions be supra-optimal. Thus, the transient rise in root elongation which sometimes occurs following excision of the tip can be envisaged as being due to a fall in auxin level within the root subsequent to removal of its site of synthesis, so that auxin is no longer supra-optimal for elongation. The high degree of sensitivity of root tissues to auxin will be referred to later in considering tropisms (Chapter 7).

It has been suggested that auxin inhibition of root elongation is mediated through an effect on ethylene production (see p. 78). However, it has been found recently that whereas ethylene inhibition of elongation in excised pea root tips is almost totally irreversible (i.e. removal of ethylene does not allow growth to resume), inhibition by IAA is fully reversible even after 16 hours of treatment. Also, although IAA inhibits root elongation immediately it is applied, ethylene is generally without effect for the first 3 to 6 hours of treatment. These findings suggest that inhibition of root elongation by auxin is not mediated through an effect on ethylene synthesis.

A positive correlation has been found between endogenous auxin concentrations and growth rate in different regions of the root of *Lens culinaris* (lentil). This constitutes circumstantial evidence that auxin is a controlling factor in root elongation, in the same way as similar evidence referred to above for coleoptile and stem elongation.

A role for gibberellins in root growth has not yet been established. Applications of gibberellins to plants generally have little effect upon the growth of the roots, except that at high dosage rates they sometimes inhibit root extension growth. Nevertheless, it has been reported that elongation of lettuce seedling radicles is stimulated by gibberellin treatment, and it is known that root tips normally synthesize gibberellins. It seems likely, therefore, that endogenous gibberellins are involved in the control of root growth and differentiation, though much more research needs to be done to establish this.

Our knowledge of cytokinins and their possible functions in root growth is extremely limited. Treatment of excised roots growing in culture with a cytokinin, in combination with an auxin, results in a stimulated cell division rate. However, this usually does not lead to an increased rate of root elongation, but to a stimulation of division of cells which are destined to differentiate into vascular tissues. Thus, it appears likely that endogenous cytokinins are involved in the control of vascular development in the root, but it is not possible at present to determine any function for cytokinins in root elongation growth.

HORMONES AND THE GROWTH OF LEAVES

Once a leaf primordium has been initiated at the stem apex, it starts to grow and develop by the processes of cell division, cell enlargement and differentiation (see Chapter 2). One can reasonably assume that these processes are under the controlling influence of growth hormones, one of which would be expected to be auxin. However, it cannot be said that auxin is involved in all aspects of leaf growth. It has been found that auxins will, depending on their concentration, either stimulate or inhibit the growth of midrib and veins but have little effect on the interveinal mesophyll tissues. At the present time little is known of the hormonal control of leaf growth, other than that auxin appears to be necessary for vein growth.

It has been suggested that a growth hormone synthesized in the roots controls mesophyll growth. Thus, it is found that if young root tips are cut off as fast as they are formed in, for example, the horseradish plant (*Amoracea lapathifolia*), normal development of the interveinal tissue fails to occur (Fig. 53). Recent evidence that roots export cytokinins and gibberellins to the shoot may provide a partial explanation for failure of mesophyll growth in such root-pruning experiments.

It has been found that treatment of some plant species (e.g. *Triticum* and *Phaseolus*) with gibberellic acid leads to a stimulation of leaf growth. Other plant species (e.g. *Pisum sativum*) do not respond in this way, and in fact mesophyll growth may be retarded following gibberellin treatment. In those species where gibberellin treatment does stimulate leaf growth, the mesophyll and vein tissues respond nearly equally. Similarly, excised leaves or leaf disks will, if

FIG. 53. Effect of repeated excision of root tips upon the growth of mesophyll in leaves of the horseradish plant (*Amoracea lapathifolia*). The smallest and most deeply lobed leaves are the youngest. (Photographed by Mr. J. Champion, from a plant grown by Dr. J. Dore.)

floated on the surface of a solution of gibberellin or cytokinin, often expand due to growth of the mesophyll. Thus, generally speaking, gibberellins and cytokinins differ from auxins in their effects on leaf growth, in that the former two classes of hormone can promote mesophyll growth whereas auxins do not. This immediately suggests that endogenous gibberellins and cytokinins are important regulators of the growth of leaves. In fact it has been found that gibberellins are normally present in leaves, and that the quantity present is closely related to the growth rate of the leaf, so that young, rapidly growing leaves contain more gibberellin than do older leaves (Fig. 54). Natural gibberellins may, therefore, be of importance in the control of mesophyll growth. We know little about endogenous cytokinins in leaves, although some experiments have shown that synthetic cytokinins can replace the need for a root system for healthy leaf growth.

The growth of leaves may, therefore, be under the controlling influence of auxins, gibberellins and cytokinins, together with nutrients and perhaps other unknown hormonal factors, but it should be borne in mind that current knowledge is meagre and fragmentary.

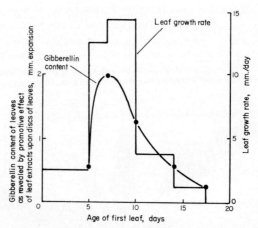

FIG. 54. Relationship between growth rate (histogram) and gibberellin content (graph) in first leaf of dwarf French bean (*Phaseolus vulgaris*). Maximum gibberellin content of the leaves occurred at seven days of age, coinciding with the period of most vigorous leaf expansion. (From A. W. Wheeler, *J. Exp. Bot.* **11**, 217–26, 1960.)

GROWTH HORMONES AND CAMBIAL ACTIVITY

We saw earlier (p. 42) that developing buds and leaves exert a stimulating effect on the development of vascular tissue in the internodes below. Thus, evidently young buds and leaves produce some kind of stimulus which promotes the formation of vascular tissue, and since they are centres of production of auxin and gibberellin, it is possible that these hormones constitute the stimulus. Rather strong evidence for this hypothesis is provided by studies on the regeneration of vascular tissue which will be described later (p. 157), but there is good evidence that the formation of secondary xylem and phloem is also under hormonal control, and this will now be considered.

The aspects of growth so far considered in this chapter as being under the control of auxins are predominantly reflections of the stimulatory effect of auxins on cell enlargement. There are, however, known instances where auxins stimulate cells to divide. A very good example of this is seen in the effect of auxins on the division of cambium leading to the formation of xylem and phloem vascular tissues. The first realization that auxin does stimulate the cambium to divide came from experiments performed by Snow in 1935 with sunflower (*Helianthus annuus*) plants which had been "decapitated" (i.e. the apical bud was excised). In such decapitated plants, it was found that the fascicular cambium of young internodes failed to divide, and also that inter-fascicular cambium did not form. In other words, excision of the apical bud prevented the formation of secondary vascular tissues (xylem and phloem).

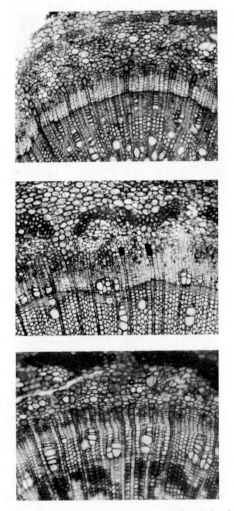

FIG. 55A. Effect of auxin and gibberellin on cambial activity in poplar (*Populus robusta*) twigs. *Top*: twig treated with gibberellic acid (GA₃); *centre*: treated with indole-3-acetic acid (IAA); *below*: treated with IAA and GA₃ in combination.

However, when IAA was applied to the upper cut end of the stem in decapitated plants, then normal cambial activity and secondary thickening resulted, suggesting that cambial activity depends upon the supply of auxin produced in the apical bud and moving basipetally in the stem.

There is considerable evidence that the initiation of cambial activity in deciduous trees in the spring occurs in response to auxin arising in the expanding buds. For example, cambial divisions commence immediately below the expanding buds, and then spread *downwards*, i.e. basipetally, through the twigs

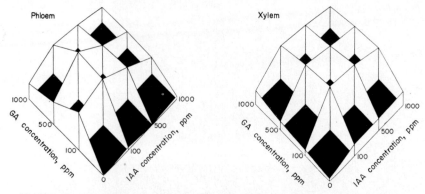

Fig 55B. Quantitative effects of IAA and GA$_3$ on cambial activity of poplar. The two hormones were applied in lanolin in various combinations at the concentrations shown. Vertical axis indicates width of new xylem and phloem in eye-piece units. See text for explanation. (From P. F. Wareing, C. Hanney and J. Digby, in *The Formation of Wood in Forest Trees*, Academic Press, New York, pp. 323–44, 1964.)

to the branches and trunk, but the acropetal spread of cambial activity is never observed. Thus, there is a parallel between the pattern of cambial initiation and the polar movement of auxin in stems. Disbudding twigs in the spring prevents the onset of cell divisions in the cambium, but application of auxin to the upper end of a disbudded twig allows normal basipetal spread of cambial activity to occur.

Gibberellins also have an effect on cambial activity, as was first shown by spraying leafy shoots of apricot trees with GA$_3$ which caused increased cambial division and xylem development.

The effects of hormones on cambial activity can conveniently be studied by taking pieces of stem of woody species in the early spring, before bud expansion has commenced, removing the buds, and applying the hormones, either mixed with lanolin or in aqueous solution, to the apical ends of the stem segments. After about 2 weeks the stems are sectioned for observations on cambial activity. With no applied hormone there is no cambial division, but if IAA is applied there is limited cambial division and some differentiation of new xylem elements can be observed (Fig. 55A). If GA$_3$ alone is applied, cambial division occurs, but the derivative cells on the inner (xylem) side remain undifferentiated and retain their protoplasmic contents; however, careful observation shows that some new phloem, with differentiated sieve tubes, is formed in response to GA$_3$. When IAA and GA$_3$ are applied together, there is greatly increased cambial division and normal, differentiated xylem and phloem are formed. By measuring the width of the new xylem and phloem it is possible to study the interaction of IAA and GA$_3$ quantitatively (Figure 55B).

Little is known of the role of cytokinins in normal cambial activity, but studies with isolated pea stem sections have indicated that these hormones can also stimulate cambial division and increase the lignification of the developing xylem cells. It is probable, therefore, that normal secondary thickening in stems involves auxins, gibberellins and, perhaps, cytokinins.

FURTHER READING

General

AUDUS, L. J. *Plant Growth Substances*, 2nd ed., L. Hill Ltd., London, 1959.

LAETSCH, W. M. and R. E. CLELAND, (Ed.). *Papers on Plant Growth and Development*. Little, Brown & Co., Boston, 1967.

LEOPOLD, A. C. *Plant Growth and Development*, McGraw-Hill, New York, 1964.

WILKINS, M. B. (Ed.). *Physiology of Plant Growth and Development*, McGraw-Hill, London, 1969.

More Advanced Reading

BRIAN, P. W. The gibberellins as hormones. *Internat. Rev. Cytology*, **19**, 229, 1966.

HESLOP-HARRISON, J. Plant growth substances, in *Vistas in Botany*, Vol. 3 (Ed. W. B. Turrill), Pergamon Press, Oxford, 1963.

PALEG, L. G. Physiological effects of gibberellins. *Ann. Rev. Plant Physiol.* **16**, 291, 1965.

SACHS, R. Stem elongation. *Ann. Rev. Plant Physiol.* **16,** 73, 1965.

STREET, H. E. The physiology of root growth. *Ann. Rev. Plant Physiol.* **17,** 315, 1966.

WAREING, P. F., C. E. A. HANNEY and J. DIGBY, The role of endogenous hormones in cambial activity and xylem differentiation, in *The Formation of Wood in Forest Trees* (Ed. M. Zimmerman), Academic Press, New York, 1964.

WIGHTMAN, F. and G. SETTERFIELD (Eds.) *The Biochemistry and Physiology of Plant Growth Substances*, The Runge Press, Ottawa, 1969.

CHAPTER 6

Other Aspects of Hormonal Control

IN THE preceding chapter we considered the role of hormones as controlling factors in shoot and root growth. However, there is good reason for believing that hormones play an important role in several other aspects of development, and we shall now consider their possible role in fruit growth, abscission and apical dominance.

HORMONES AND FRUIT GROWTH

Probably because of the economic importance of fruits, the physiology of their growth, development and ripening has been very intensively studied. Most botany textbooks classify "fruits" according to a rigid and complicated system based upon morphological characters, but J. P. Nitsch has pointed out that a fruit is a physiological rather than a morphological entity. To the physiologist, and incidentally the layman, a "fruit" is simply a structure which arises by development of tissues which support the ovules. Nitsch has pointed out that such a view is valid even for seedless fruits because ovules were initially present in them.

The early growth of the ovary, which occurs during development of the flower, involves cell division but little cell vacuolation. In many species, cell division ceases at or shortly after anthesis (flower opening) and the subsequent growth of the fruit following pollination is primarily due to an increase in cell size rather than in cell number. For example, in tomato (*Lycopersicum esculentum*) and blackcurrant (*Ribes nigrum*), cell division ceases at anthesis and the whole of the subsequent growth is due to cell expansion. In such species, the final size will be a function of the number of cells already present in the ovary at anthesis. On the other hand, in some species (e.g. apple) cell division may continue for a time after pollination has occurred.

Fruit cells may enlarge by vacuolation to relatively enormous sizes, and mature fruits of water melon (*Citrullus vulgaris*) consist mainly of cells so large that they are distinguishable individually by the naked eye.

Fruit Set

The early development of the ovule and ovary takes place along with other aspects of flower development (p. 46). In some species, the ovary ceases growth at the time of, or before, anthesis (Fig. 56). In others, growth goes on for a time after anthesis and prior to pollination. In both cases, further growth of the ovary takes places only if pollination is effected. Should, for some reason, pollination not occur, then growth and development of the fruit ceases. Failure of the pollination mechanism usually results in the shedding from the plant of the unfertilized flower, often mediated by the formation of a separation, or abscission layer (see p. 118) in the peduncle. Successful pollination, on the

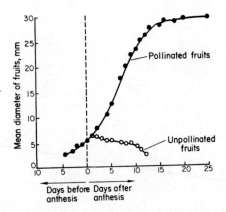

Fɪɢ. 56. Growth curves of ovary of *Cucumis anguria*. Growth in unpollinated ovaries ceases soon after anthesis, but pollinated ovaries show a typical sigmoid growth curve. The decrease in ovary diameter seen in unpollinated ovaries is due to "shrivelling". (From J. P. Nitsch, *Quarterly Rev. Biol.* **27**, 33–57, 1952.)

other hand, is followed by rapid growth of the ovary and fruit development begins (Fig. 56). At the same time, the petals and stamens wither and often abscind. The start of ovule growth and withering of stamens and petals marks the start of fruit development, and this phase is often called *fruit set*. The process of pollination, whether or not it is followed by fertilization, is apparently sufficient to cause an initial stimulation of growth in the ovary and other parts of the future fruit. This is shown by the fact that in many fleshy fruits the increase in ovary growth may start before there has been sufficient time for fertilization to occur. Moreover, even "foreign" pollen, derived from an unrelated species and hence unable to effect fertilization, may cause marked stimulation of ovary growth.

The stimulatory effect of pollen on ovary growth appears to be due to the auxin which it contains. In 1909, Fitting observed that water extracts of orchid pollen were able to induce swelling of unfertilized orchid ovaries and withering of the petals, and it was shown later that pollen is a rich source of auxin. Finally, it was shown that pure preparations of IAA would, if applied to unfertilized flowers of a number of species, induce fruit set in the absence of any pollen.

Fig. 57. Correlation between achene development and receptacle growth in straw-berry. A: only one fertilized achene present; B; three fertilized achenes present. C: *left*, control fruit; *right*, fruit with three vertical rows of achenes. D: *left*, fruit with two rows of achenes; *right*, control fruit. E: three strawberries of the same age; *left*, control; *middle*, all achenes removed and receptacle smeared with lanolin paste; *right*, all achenes removed and receptacle smeared with lanolin paste containing 100 ppm of the auxin β-naphthoxyacetic acid. (Original prints kindly supplied by Dr. J. P. Nitsch, Laboratoire du Phytotron, Gif-sur-Yvette, France. From *Amer. J. Bot.* **37**, 211–15, 1950.)

Among the species in which fruits can be set by auxins are tomato, pepper, tobacco, holly, figs and blackberry. In these species, fruits which have been set by treating unpollinated flowers are seedless. The production of such seedless fruits is called *parthenocarpy*. Recently it has been found that ethylene will simu-late the effects of pollen on ovary swelling and petal withering in orchids.

Thus, it is possible that an auxin effect is mediated through an influence on ethylene production.

Fruit Growth

Although pollination may stimulate the initial swelling of the fruit, in most species further development of the fruit appears to be dependent upon the presence of developing seeds, and hence can only occur when fertilization is effected. Thus, in many fruits, such as grapes, blackcurrants, tomatoes, apples and pears, strong correlations exist between the final size of the fruit and the number of fully developed seeds it contains. In the case of the strawberry, Nitsch showed by elegant surgical experiments that receptacle growth is dependent upon achene development (Fig. 57).

The interaction between the growth of the ovary and that of the embryo

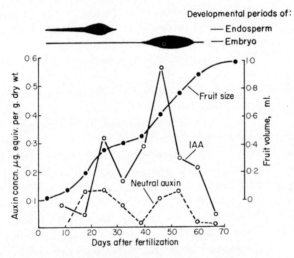

FIG. 58. Changes in concentration of indole-3-acetic acid (IAA) and an unidentified neutral auxin in the blackcurrant berry in relation to the double sigmoid growth curve of the berry and the main developmental periods of the endosperm and embryo. (From S. T. C. Wright, *J. Hort. Sci.* **31**, 196, 1956.)

and endosperm is shown by the changing growth rates of these different parts of the fruit at different stages of development. In some instances the growth curve for the fruit is sigmoid (e.g. in apple) and in others it is doubly sigmoid (Fig. 58). In peach, the changes in growth rate of the pericarp can apparently be correlated with changes in growth rate of the developing seeds. The stimulatory effect of the developing seeds upon the growth of pericarp tissue appears to be due, at least partly, to the auxin which they produce. Developing seeds are rich sources of auxin and it has been shown that there is a declining gradient of auxin concentration within the tissues of the fruit, with the highest concentration in the seeds, less in the placenta and least in the carpel wall. This gradient is consistent with the view that auxin is produced in the developing seeds and moves outwards from there to the other parts of the fruit.

A good example of the relationships that have been found between endogenous auxins and fruit growth has come from studies of berries of the blackcurrant, which show a double sigmoid growth curve (Fig. 58). Two auxins were found in the berries, one acidic (possible IAA) and the other neutral (possibly IAN). These two auxins were found to be produced mainly at the times of endosperm and embryo development, which in turn coincided with the times of maximal growth rate of the fruit. It appears likely, therefore, that the first grand period of fruit growth in blackcurrant is under the control of auxins produced in the developing endosperm, and that the second grand period is induced by auxin originating in the growing embryo. Similar patterns of auxin production in relation to fruit growth have been reported for other species by several workers, but there are other conflicting reports that no direct correlation occurs in some species between auxin content and fruit growth rates.

It is important to realize that despite the evidence implicating auxins in flower and fruit growth, there is also a considerable likelihood that auxins are not the only hormones involved. It has proved impossible with many plant species to induce parthenocarpic fruit development by auxin treatment, but it has proved possible to do this by spraying the flowers with a solution of gibberellin (e.g. in members of the genus *Prunus* such as cherries, peaches and almonds). However, indicative as the effects of applied gibberellins on fruit development may be, this does not in itself provide incontrovertible evidence that internally produced gibberellins are concerned in the normal growth of fruits and seeds. Consequently, studies have been made of the gibberellin contents of various fruits and seeds at different stages of their development, and in general it has been found that young developing seeds contain relatively large amounts of gibberellins. As the seeds mature and their growth slows down there is a simultaneous fall in their gibberellin content. It appears likely that gibberellins move out from the young developing seeds in a similar manner to that suggested for auxin, and that both types of hormone are involved in the control of fruit growth.

The third type of growth hormone likely to be participating in the processes of growth in a fruit are the cytokinins. As discussed earlier (Chapter 4), cytokinins are growth hormones particularly concerned in the control of cell division, and it is, therefore, likely that the active cell division which is known to occur in young developing fruits is under the controlling influence of this type of hormone. Evidence that cytokinins are involved in fruit growth is afforded by experiments which have shown the presence of cytokinins in the young fruits of apple, banana and tomato during the stages of growth in which cell division is most rapid. Thus, we may regard fruit growth and development as being under the control of several types of growth hormone, a situation which probably obtains in all phases of plant growth.

Parthenocarpy

We have already seen that in some species parthenocarpic fruits may be produced by treating unpollinated flowers with auxins or gibberellins. In addition to the experimental production of such fruits, natural parthenocarpy may occur in certain species. Thus, horticultural varieties of bananas, pine-apples, cucumbers, tomatoes and figs exist, in which seedless fruits are normally produced without the need for any exogenous hormone. In some species, fruits are formed without pollination, while in others pollination is necessary but fertilization does not occur; in others again, fertilization occurs but the embryos abort before the fruit matures. It is not known how the growth of these seedless fruits is controlled, but it seems possible that in some cases the maternal tissues, such as the placenta, may be capable of producing auxin in the absence of normal embryos. Thus, it has been shown that the ovaries of unopened flowers of parthenocarpic varieties of orange and grape have a higher auxin content than do those of normal seeded varieties. Moreover, it has been found that young parthenocarpic fruits of cucumber contain seed-like structures, but which lack embryos and endosperm, and it is possible that these are centres of auxin production.

The production of fully developed fruits by treatment with a single applica-tion of auxin to the flowers (p. 114) also poses a number of problems, since it is not to be expected that the auxin which is applied in this way will itself be sufficient to supply the needs of the developing fruits over several weeks. How-ever, it has been found that pollination stimulates the production of auxin by the tissues of the ovary itself in some species, such as tobacco, and it is possible that external application of auxin may similarly trigger off the produc-tion of auxin by certain tissues of the fruit, and that once this has occurred, the production of endogenous auxin will meet the requirements for the further development of the fruit. Application of auxin has been shown to lead to growth of unfertilized ovules, which develop normal looking seed coats but which contain no embryos. It is interesting to note that in some species, such as olive, hops and maize, application of auxin will stimulate initial fruit set, but further development of the fruit does not occur without pollination. Possibly, in such species the exogenous auxin does not stimulate the production of endogenous auxin necessary for further fruit development.

HORMONES AND ABSCISSION

Leaf Abscission

In most plant species there comes a time during the life of each leaf when it is shed from the stem. This occurs most obviously at the end of the growing season in temperate regions of the world, but leaf fall is not confined to autumn. All through the summer in temperate areas, and all through the year in the tropics, there is a less conspicuous but continuous dropping of older leaves from plants. The process by which leaves (and also other organs such as fruits and flowers) are removed from the plant is known as *abscission*. The act of abscission

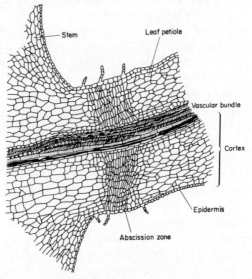

FIG. 59. Longitudinal section through the abscission region of the base of the petiole of a leaf of a typical dicotpledonous plant. (From J. Torrey, *Development in Flowering Plants*, MacMillan, New York, 1967.)

of an organ, such as a leaf, is usually achieved by the formation of a *separation* or *abscission layer* at the base of the petiole. This is a thin plate of cells oriented at right angles to the axis of the petiole (Fig. 59). The walls of the separation layer cells become softened and gelatinous, so forming a weak region, which readily breaks under the strain of wind-induced movement of the leaf. Softening of the cell walls in the separation layer is a consequence of increased methylation of pectic acid to yield pectin (p. 93). Separation layers of a similar nature are also formed in the stalks of flowers and fruits.

It is known that the formation of a separation layer in the petiole is related to a fall in the auxin content of the leaf. As a leaf gets older its auxin content declines, eventually to the point where it contains only as much as or perhaps less than the tissues of the stem. When that happens the separation layer forms, although precisely why it should is not yet fully understood.

One theory is that there is an auxin concentration gradient across the abscission zone, between stem and petiole, which determines whether or not a separation layer is formed. It has been observed that an application of an auxin to the distal cut end of a petiole after removal of the auxin-synthesizing lamina, delays abscission of the petiole. Conversely, if the auxin is supplied not to the distal cut end of the petiole, but to the proximal (stem) side of the abscission zone, then frequently one sees that abscission is accelerated rather than delayed by auxin. However, Jacobs has pointed out that all experiments showing acceleration of abscission by proximal applications of auxin have been conducted on excised nodal segments kept under non-sterile conditions. Under sterile conditions, petiole abscission in *Coleus* nodal segments was always delayed by distal or proximal applications of IAA. Jacobs suggests that some instances of auxins accelerating abscission were due to artifacts in the experimental system, and that provided auxin is applied immediately after removal of the lamina, and the abscission zone is both free of contaminants and well supplied with nutrients, then auxin invariably delays abscission. Nevertheless, there is good evidence that auxin *can* accelerate abscission if it is applied to the distal end of a petiole some hours after excision of the lamina. Because of this, it has been suggested that the abscission process can be divided into two phases—an initial one ("stage 1") inhibited by auxin, and a second one ("stage 2") which may actually be promoted by auxin. The second stage is also stimulated by other substances, such as the amino acid alanine, and ethylene.

The latter substance appears to play a very important role in the induction of abscission, for it is the most potent chemical known which can cause leaf fall. Ethylene does not, however, appear to be very effective in bringing about leaf abscission unless the auxin content of the leaf is low. Thus, old leaves (low in auxin) are more likely to abscind (= absciss) in the presence of ethylene than are young auxin-rich leaves. Similarly, treatment of debladed petioles with auxin delays the transition from "stage 1" to the ethylene-sensitive "stage 2" of the abscission process. The change over from the stage 1 to the stage 2 condition is in some way a consequence of senescence in the petiole. Thus, petioles in the stage 2 condition have undergone a measure of senescence which makes them more sensitive to ethylene. The reasons for the propensity of senescent petioles to abscind in the presence of ethylene are not clear, but one factor is that senescent tissues respond to ethylene treatment by synthesizing even more ethylene, whereas this does not occur to such a marked degree in petioles in the stage 1 state. The principal effect of auxin in delaying abscission is, therefore, to prolong stage 1 (i.e. to delay senescence, which in turn maintains relative insensitivity to ethylene so far as abscission is concerned).

In addition to the delaying effect of auxin, and the promoting influence of ethylene, in leaf abscission, there is also evidence for the existence of other factors concerned in the control of abscission. Thus, there is a suggestion that a "senescence factor" is produced in yellowing leaves which promotes separation layer formation. The identity of this senescence factor has never been satisfactorily determined, but it might be abscisic acid (Chapter 11), or amino acids such as alanine being retranslocated from senescing tissues of the lamina and petiole. However, it is not clear whether abscisic acid is directly involved in the normal control of leaf abscission or whether it acts indirectly by promoting leaf senescence.

Fruit Abscission

Abscission phenomena are also shown by flowers and fruits. Thus, an abscission zone is commonly found at the base of the pedicel of the flowers in many

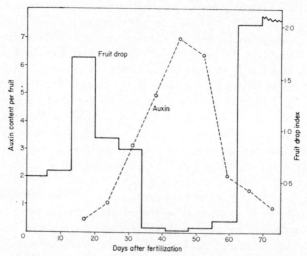

Fig. 60. Changes in the content of an unknown acidic auxin in the blackcurrant berry, in relation to the rate of fruit drop. (From S. T. C. Wright, *J. Hort. Sci.* **31**, 196, 1956.)

species, if pollination and fertilization fail to take place. Similarly, even when successful fertilization has occurred, an abscission zone, leading to fruit-drop, may be formed at various stages during the development of the fruit. This is well seen in certain varieties of apple, in which there may be three peak periods of fruit drop: (1) immediately after pollination ("post-blossom" drop), (2) soon after growth of the young fruits ("June drop"), and (3) during ripening ("pre-harvest drop").

For some species it has been shown that the periods of fruit drop coincide with periods of low auxin content in the fruit and conversely the times of low drop rate occur when the auxin content is high. This situation is found in the blackcurrant, in which we have already seen that it is possible to correlate growth rate with the levels of two auxins, one acidic and one neutral (Fig. 58). It has also been shown that there is a second acidic auxin the levels of which show a different pattern of variation which correlates inversely with the rate of fruit abscission (Fig. 60). This situation is analogous to that in leaves, therefore, where the formation of an abscission layer is associated with diminished auxin levels in the lamina.

On the other hand, in other species it has not been possible to correlate the variations in rate of fruit drop with auxin levels and it seems likely that in such cases other factors play a role in determining fruit drop. Indeed, there is little doubt that ethylene is involved in fruit, as well as leaf, abscission, for this substance promotes fruit senescence and abscission in several species.

GROWTH CORRELATION

The various growth processes that proceed simultaneously in a plant are not independent, but are closely linked with one another. Thus, as a stem increases in length by the activities of the apical meristem, its strength is increased by activities in the cambium leading to increased girth and rigidity of the older parts, so enabling the whole shoot to stand erect. Further, as the shoot increases in size, so does the need for water and mineral nutrients increase, and this is catered for by a nicely balanced relationship between shoot growth and root growth. As the shoot increases in bulk, the size of the root system becomes proportionately larger, which allows for the additional mineral nutritional requirements of the shoot to be met. To some extent these *growth correlations*, as they are called, are explicable in terms of the availability of nutrients, or food factors, and the competition between growing regions for these substances. Thus, shoot and root growth are related to one another, and this is probably due in part to their mutual nutritional dependence; the shoot supplies the organic material which the root is unable to manufacture for itself, and in return obtains from the root the water and mineral salts to which it does not itself have direct access. In the same way, vegetative growth is very reduced when a plant is fruiting, probably mainly as a consequence of the diversion of the available food materials into the developing seeds and fruits (see Chapter 12).

This is certainly not the whole story, however. Competition for available nutrients does not adequately explain why active growth at any one time is usually restricted to only a few of the many places in a plant where it is potentially possible. An example of this is the fact that the apical meristem of a shoot is usually in a state of active growth, whereas growth of the axillary buds below it is often strongly inhibited. This characteristic of shoot growth is known as *apical dominance* or *correlative inhibition* of buds. A related phenomenon is seen in the inhibiting effect of the main root apex upon the initiation of lateral roots in the pericycle cells. Removal of the root apex causes an increase in lateral root formation in the remainder of the root, a result analogous to the effect of removing the shoot apical bud upon the growth of axillary buds. However, the mechanism by which these phenomena occur is not necessarily the same in shoot and root.

It is known that growth hormones play important roles in the correlation of growth in different parts of the plant. It is likely that all growth correlations are in one way or another affected by patterns of hormone distribution within the plant. We have already seen several examples of correlative effects in plant growth which are mediated by growth hormones, as in the stimulation of cambial activity by auxin and gibberellins arising in the buds of woody shoots, and the stimulation of fruit growth by hormones produced by the developing seeds. We shall now consider the role of hormones in apical dominance.

APICAL DOMINANCE

The apical bud of a shoot usually grows much more vigorously than the axillary buds, despite the fact that it is apparently the least favourably situated bud in relation to the supply of organic and inorganic nutrients from the mature leaves and the root system. There is a great deal of variability between species with respect to the degree of dominance of the apical bud over the lower axillary buds. In some species, such as tall varieties of sunflower (*Helianthus annuus*), the dominance is complete and extends over almost the whole length of the stem. In others, such as tomato, the dominance of the apical bud is weaker, and the axillary buds situated only a little way below the main shoot tip grow out, resulting in a bushy shoot system.

In many species the dominance of the shoot tip becomes weaker as the plant gets older. This is seen clearly in plants such as sycamore (*Acer pseudoplatanus*) or ash (*Fraxinus excelsior*) where the early years of growth are characterized by strong growth of the leading shoot, whereas in later years a branching habit is seen. Even in herbaceous annuals there is often a weakening of apical dominance towards the end of the growing season, and in those species where the apical meristem eventually changes to produce a terminal flower, this often coincides with a release of axillary buds from correlative inhibition.

If a shoot is decapitated, i.e. the apical bud cut off, then one or more of the lower axillary buds grow out. Usually one of the outgrowing laterals becomes dominant over the others, exerting an inhibitory influence on their growth. Where this happens it is frequently the uppermost of the axillary buds which becomes the dominant shoot.

The first demonstration that auxin is involved in apical dominance phenomena came in 1934. At that time it was known that the apical bud is a principal centre of auxin production in the plant, and the plant physiologists Thimann and Skoog considered it possible that it is auxin, synthesized in the apical bud, which causes the lateral buds to be suppressed. These workers, therefore, carried out an experiment in which they compared the outgrowth of axillary buds in three sets of bean plants (*Vicia faba*)—a set of intact control plants, a second set which had been decapitated, and a third set which had been

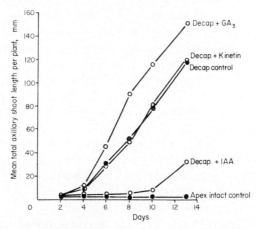

Fig. 61. Effect of hormones on outgrowth of axillary buds of pea plants (*Pisum sativum*) following decapitation. The hormones referred to were applied in lanolin (concentration 1,000 ppm) to the decapitated stumps of the plants. The ordinates show the total growth of the three remaining axillary buds per pea plant. (I. D. J. Phillips, original data.)

decapitated but to which a block of agar containing the auxin IAA had been placed on the cut surface of the stem. They found that the lateral buds grew out vigorously in the decapitated plants, but that treatment with IAA prevented this outgrowth. They concluded, therefore, that it was the auxin produced by the apical bud which caused the inhibition of the lateral buds. Since Thimann and Skoog's first report, there have been many other demonstrations that added auxin, when substituted for the apical bud, can inhibit the growth of the lateral shoots in many plant species (Fig. 61). There is, however, a certain amount of controversy as to whether IAA can substitute for the apical bud in certain species, such as *Coleus blumei*.

The way in which auxin causes the inhibition of axillary bud growth is not, however, fully understood at the present time. It appears paradoxical that auxin, which we have so far been considering as a promoter of growth, should cause an inhibition of growth in axillary buds. To resolve this contradiction it was suggested by Thimann in 1937, that the optional auxin concentration for bud growth is lower than for stems (Fig. 51) and that the bud is inhibited by the "supra-optimal" auxin concentration normally present in the stem, as a result of its synthesis in the young expanding leaves, and basipetal transport from there. This is known as the "Direct Theory" of auxin inhibition of lateral buds. It is, of course, based on the assumption that auxin enters lateral buds from the stem.

The validity of Thimann's direct theory is considered doubtful today. One of the principal objections to the theory is that determinations of the auxin contents of inhibited lateral buds in *Lupinus*, *Pisum sativum* and *Syringa vulgaris* have revealed that the auxin levels, far from being supra-optimal, are, in fact, sub-optimal. Also, it has been found that applications of low concentrations of auxin to the stumps of decapitated *Lupinus* and *Phaseolus multiflorus* plants actually accelerated growth of laterals, and that only at higher auxin concentrations did inhibition occur. Consequently, it is now generally considered that auxin does not exert its inhibitory effect on lateral buds in such a direct manner as that originally proposed by Thimann.

One of the early hypotheses for apical dominance assumed that since the apical meristem is the first-formed shoot meristem in the germinating seedlings, then it would continue to command a preferential supply of metabolites as these moved along their concentration gradients. This was known as the "Nutritive Theory" of apical dominance.

If the nutritive theory is correct, one might expect that the dominance of the apical bud over the lateral buds to be most clearly manifest when a plant is deficient in nutrients, for example, when growing in soil low in necessary mineral elements. This has been clearly shown to be the case for several plants, particularly flax (*Linum usitatissimum*) in which lateral growth is entirely suppressed by the terminal bud under conditions of mineral nutrient deficiency, whereas lateral growth occurs freely under conditions of high mineral (particularly nitrogen) nutrition. One can assume that when nutrients are freely available to the plant, there are sufficient "left over" after the apical bud has received its necessary quota to allow movement of nutrients into the lateral buds. The effect of the apical bud in flax plants growing under high or low nutrient conditions was exactly duplicated by IAA applications to the stump of decapitated plants.

Although it is undoubtedly true that the apical bud does obtain more available nutrients than the axillary buds, the simple "source-sink" explanation of the nutritive theory does not adequately explain why auxin can substitute for the apical bud in correlative inhibition of axillary buds. Further, many studies with isotopes such as ^{32}P-phosphate and ^{14}C-sucrose have demonstrated that nutrients do indeed move to, and accumulate in regions of high exogenous auxin concentration (Fig. 62). This "auxin-directed" transport of metabolites indicates that auxin production in the apical bud, and its basipetal transport, induces movement of available nutrients towards the region of highest auxin concentration, i.e. to the apical bud itself. It is not clear how this comes about, but there is evidence that basipetally moving auxin in the stem inhibits the development of vascular connections between axillary bud and stele, so reducing

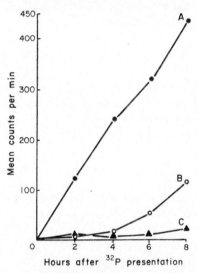

FIG. 62. Auxin-directed transport. The effect of IAA on the rate of accumulation of ^{32}P at the top of the stem in decapitated pea plants is shown. The radioactive ^{32}P was applied to the base of the stem just above soil level. In decapitated plants not supplied with auxin very little accumulation of ^{32}P occurred at the top of the stem (C). Application of IAA to the top of the stem immediately after excision of the apical bud (A) greatly enhanced ^{32}P accumulation. Curve B shows the effect of applying the auxin six hours after decapitation. (Reprinted from C. R. Davies and P. F. Wareing, *Planta*, **65,** 139–56, 1965.)

the capacity of the axillary buds to obtain a supply of nutrients *via* the vascular system. Decapitation of the shoot lowers the auxin level of the stem with a resultant rapid development of bud-stem vascular connections.

Recent research has indicated that cytokinins may also play some part, along with auxins, in the hormonal regulation of apical dominance. For example, it has been found that direct application of a cytokinin to a correlatively inhibited axillary bud can release that bud from inhibition. However, in all cases of bud release by cytokinin it has been found that growth of the axillary shoot continued for only a short time, after which apical dominance was reimposed. Only when auxin was supplied directly to outgrowing cyto-kinin-treated buds did growth continue. Consequently, it appears that sufficient cytokinin and auxin have to be present before a bud is able to grow (this observation provides further evidence that the auxin concentration in inhibited buds is sub- rather than supra-optimal). We know from other research that cytokinins are exported from the root to the shoot (p. 265) and it is possible that it is the availability of these endogenous cytokinins to axillary buds which controls their outgrowth in at least the initial stages.

Available evidence suggests that gibberellins may not be directly involved in the control of apical dominance phenomena in the way that auxins, and perhaps cytokinins are. Thus, gibberellins do not substitute for the apical bud in the correlative inhibition of axillary buds (Fig. 61), nor do they cause the release of buds from correlative inhibition (cytokinins can do so). Application of exogenous gibberellin to an intact plant often results in an increase in apical dominance, but this is probably due to effects on internal auxin level and distribution in the gibberellin-treated plants.

Hormonal Interaction in Stolon Development

A striking example of the importance of hormonal interaction in the control of shoot growth is seen in stolon development in the potato plant. The stolons are axillary shoots which normally arise from the basal nodes of the stem below the soil surface. They differ from the erect, leafy axillary shoots which arise on the aerial part of the shoot in that they have (1) only rudimentary leaves, (2) elongated internodes, and (3) poor development of chlorophyll (even in the light), and (4) they grow horizontally. Apical dominance evidently plays an important role in stolon development, since if the aerial shoot is decapitated and all axillary shoots are removed from it, the stolons will turn up and produce normal-looking, erect leafy shoots.

Stolons are not normally formed on the upper, leafy part of the aerial shoot, but they may be induced to develop by decapitating the shoot and applying exogenous hormones. If IAA alone is applied to the stump of a decapitated shoot, the uppermost axillary shoot is partially inhibited, whereas if GA_3 alone is applied, extension of this axillary is promoted. By contrast, when IAA and GA_3 are applied together to the decapitated stem, the development of the uppermost axillary is dramatically changed and it becomes a horizontal, stolon-like structure (Fig. 63). If kinetin alone is applied to the decapitated stem there is no observable effect, but if kinetin or benzyladenine is applied directly to the *tip* of a natural or of an experimentally promoted stolon, the latter very rapidly changes its pattern of development and turns upwards to become a leafy shoot (Fig. 63). Thus, the development of an axillary shoot of potato can be controlled very precisely by manipulating the levels of auxin, gibberellin and cytokinin. It seems likely that the natural control of stolon development involves a similar interaction between endogenous hormones.

From the various examples we have considered in this and the preceding chapter, it is evident that the correlated and integrated character of the plant

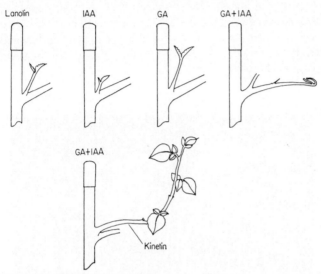

FIG. 63. Interaction between hormones in the control of stolon development in the wild potato species, *Solanum andigena*. Leafy aerial shoots were decapitated and hormones were applied in lanolin to the decapitated stems, as shown. The uppermost axillary shoot showed the following responses to hormone: control shoots (treated with plain lanolin) showed outgrowth of leafy axillary shoots; IAA alone caused some inhibition of growth of axillary; GA₃ alone caused some internode extension in the leafy axillary; IAA+GA₃ caused the axillary to grow as a horizontal, stolon-like shoot; when stolons were first stimulated to develop by application of IAA and GA₃, and then kinetin was applied to *the stolon tip*, the axillary shoot turned upwards and showed normal leaf expansion. (From A. Booth, *J. Linnean Soc.* **51**, 166, 1959, and D. Kumar and P. F. Wareing, unpublished.)

is at least partially attributable to the presence of specific amounts of growth hormones at specific places and times. There remains, however, the problem of what it is which controls the precise production and distribution of growth hormones within the plant body.

PLANT GROWTH HORMONES IN AGRICULTURE AND HORTICULTURE

The fact that the mechanism of plant growth hormone action is not understood has not prevented some of these substances becoming useful tools in agricultural and horticultural practices. The earliest commercial use of a plant hormone was the exposure of fruits to ethylene gas to accelerate ripening, a practice which is still common today. Also, we have already mentioned the usefulness of auxins and gibberellins in inducing parthenocarpic fruit development (p. 116). Another practical use for auxins is the induction of flowering, and consequently fruiting, by spraying pineapple crops with compounds such as 2,4-dichlorophenoxyacetic acid (2-4-D). The prevention of "pre-harvest drop" in apples, by synthetic auxin sprays, has also been found to have commercial value. Also, the rooting of stem cuttings is enhanced by treatment with various synthetic auxins (Fig. 64).

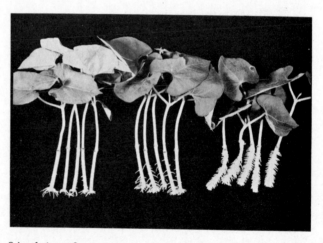

FIG. 64. Stimulation of root initiation in bean cuttings. *Left*: control cuttings (no auxin treatment); *middle*: cuttings treated with a solution of 5 mg/l naphthalene acetic acid (NAA); *right*; treated with 50 mg/lNAA. (Original print kindly supplied by Dr. L. C. Luckwill, Long Ashton Reseach Station.)

As we saw earlier (see Fig. 51), a characteristic feature of auxins is that above a certain concentration, their effect is to inhibit rather than to stimulate growth. It appears that too high an auxin concentration disorganizes the delicate machinery of growth. Plants treated with an excess of auxin become distorted, with epinastically curled leaves and split stems. They may subsequently die, but not all auxins are equally toxic, and plant species also show varying sensitivities to auxins.

The reason for toxic effects of supra-optimal auxin concentrations is unknown and is likely to remain so until an understanding is gained of the basic mechanism of auxin action. This has not, however, prevented their use in the control of weeds among crop plants. The removal of weeds by mechanical cultivation is a costly and laborious business, and the discovery that spraying a synthetic auxin, such as 2,4-D, over a field can achieve the same result has proved of enormous economic value. Of course, many chemicals other than auxins, if applied in sufficiently high concentration, are poisonous to plants and can be used as weed killers. The special merits of synthetic auxins such as 2,4-D are that they will kill plants when applied at relatively low concentrations, they are comparatively harmless to animals, they are non-corrosive, they are translocated within the plant to kill those parts, such as the roots, not reached by sprays and, most important, some auxins are *selective* and can, therefore, be used to kill weeds without damaging surrounding crop plants. As a general rule, cereals are fairly resistant to applied auxins, whereas a number of dicotyledonous weeds are very sensitive and are killed. Thus, the weed known as yellow charlock (*Sinapis arvensis*) can be controlled in cereal crops such as oat (*Avena*) or wheat (*Triticum*), and lawn-weeds such as daisies (*Bellis perennis*) and plaintains (*Plantago* spp.) can be killed without harming the grass.

The selective action of auxins as weed-killers depends on a number of factors. Very often, susceptible plants are dicotyledonous and possess broad horizontally spreading leaves, which retain auxin solution sprayed on them, while resistant plants are often monocotyledonous with narrow, erect leaves, off which the spray droplets run. In addition, the epidermis of some plants is more easily penetrated by auxin solutions than is that of others. More important than these factors, though, is the existence of inherent differences between the living cells of different species in sensitivity to synthetic auxins.

The most widely used synthetic auxins in weed control at the present time are 2,4-D, 2,4,5-T and MCPA (Fig. 41) or proprietary mixtures of these. IAA, the naturally occurring auxin of most plant species, is relatively ineffective as a weed-killer, probably due to its rapid destruction by the enzymes of the treated plants (p. 69).

Gibberellins might have at least as great a practical potential as auxins. The effectiveness of gibberellins in the induction of parthenocarpic fruits is just beginning to be realized at the commercial level, as they are often useful with species in which fruit set is not promoted by auxin. Also, gibberellins act synergistically with auxins in inducing fruit set in species such as tomato. Uses for gibberellins in the cultivation of grapes (*Vitis* spp.) include stimulation of elongation of clusters (to decrease rotting of individual berries), enlargement of berries, and increasing berry set. Strawberry plants can, by treatment with gibberellic acid, be induced to produce earlier flowers, and therefore fruits, than untreated plants. This effect can lead to commercial gain, for fresh strawberries sell for higher prices during the early part of the harvest period.

The effectiveness of gibberellins in speeding up germination rates of seeds of a large number of species, as well as in breaking dormancy (p. 235), has obvious practical possibilities.

Probably the most important single commercial application of gibberellins yet found is in the malting industry. Malting is a process whereby barley seeds are allowed to germinate for several days. The germinated seeds are then used in the preparation of a medium for fermentation yeasts in the manufacture of beer. The purpose of the malting procedure is to allow time for stored food reserves in the endosperm of each seed to be converted to other substances more suitable as substrates for yeast growth. Thus, for example, starch reserves are converted to sugars by the action of hydrolytic enzymes such as α-amylase which are formed in the aleurone cells during germination (p. 253). Due to the stimulatory effect of gibberellins on α-amylase synthesis (p. 287), treatment of barley seed with gibberellic acid both speeds up and provides more strict control, of malting. This has the benefit of conferring considerable saving of time and production of higher yields of the malting product.

It has been claimed that pasture grasses, when sprayed with gibberellic acid, grow more rapidly and are more valuable nutritionally for grazing animals. It remains to be seen whether this application will have an economic advantage.

It is highly likely that some synthetic cytokinins will be widely used in future to delay the processes of senescence in flowers and vegetables whilst these products are being transported from the grower to the consumer.

As already stated, ethylene has been used for some years for accelerating ripening processes in a number of fruits (an aspect of senescence), particularly citrus fruits such as oranges and lemons. Moreover, demonstrations that this gas is a most potent stimulator of premature leaf fall have raised the possibility of its use as a defoliant in crops such as cotton, pea, beans, etc., where mechanical harvesting of fruits is expedited if leaves are not present. However, because of its gaseous nature, ethylene does not readily lend itself to applications to outdoor crops. Because of this, a search has been made for other chemicals which promote leaf abscission. A number of such substances are now known, and the most effective of these are those which enhance ethylene synthesis in the plants to which they are applied. A recent development is the search for chemicals which can be sprayed on to plants, but which are themselves broken down in plant tissue to yield ethylene. An example of such a substance is 2-chloroethanephosphonic acid.

$$Cl.CH_2.CH_2-\overset{\overset{\displaystyle O}{\|}}{\underset{\underset{\displaystyle OH}{|}}{P}}-OH$$

2-chloroethanephosphonic acid

There is little doubt that practical uses of plant growth hormones will in the future be more extensive and varied than at present. Full realization of their potential will, however, not be achieved until our understanding of their functions and modes of action in plants is greatly increased.

FURTHER READING

General

AUDUS, L. J. *Plant Growth Substances*, 2nd. ed., L. Hill Ltd., London, 1959.
LAETSCH, W. M. and R. E. CLELAND (Ed.). *Papers on Plant Growth and Development*, Little, Brown & Co., Boston, 1967.
LEOPOLD, A. C. *Plant Growth and Development*, McGraw-Hill, New York, 1964.
WILKINS, M. B. (Ed.). *Physiology of Plant Growth and Development*, McGraw-Hill, London, 1969.

More Advanced Reading

AUDUS, L. J. Correlations, *J. Linn. Soc. (Bot.)*, **56,** 177, 1959.
CARNS, H. R. Abscission and its control. *Ann. Rev. Plant Physiol.* **17,** 295, 1966.
CHAMPAGNAT, P. Physiologie de la croissance et l'inhibition des bourgeons: Dominance apicale et phenomènes analogues, in *Encyc. Plant Physiol.* **15,** (1).
CRANE, J. C. Growth substances in fruit setting and development. *Ann. Rev. Plant Physiol.* **15,** 303, 1964.
LUCKWILL, L. C. Hormonal aspects of fruit development in higher plants. *Symp. Soc. Exp. Biol.* **11,** 63, 1967.
NITSCH, J. P. Plant hormones in the development of fruits. *Quart. Rev. Biol.* **27,** 33, 1952.
PHILLIPS, I. D. J. Apical dominance, in *The Physiology of Plant Growth and Development* (Ed. M. B. Wilkins), McGraw-Hill, London, 1969.

CHAPTER 7

Hormones and Growth Movements

IT IS one of the characteristic properties of living organisms that they have the ability to perceive and respond to changes in external or internal conditions. The change in external conditions is called the *stimulus* and the resulting change in the plant is called the *response*. The response may take place in various ways, but very frequently stimulation results in *movement*.

All plants have the capability of movement. In lower plants, such as many algae, fungi and bacteria, movement of the whole organism occurs. In higher plants the capability of movement is restricted to individual organs or parts of the whole organism. The process of straight extension growth itself can perhaps be regarded as such a movement, in that, for example, the root tip moves through the soil as a result of growth. Other types of movement result from *differential* growth rates; that is, the organ shows different rates of growth on opposite sides and this results in the bending of the organ in one direction. Such movements are termed *growth movements*. Not all plant movements are growth movements. Some are brought about by reversible turgor changes in tissues at the base of each leaf and there is often a specialized structure, the pulvinus, at which bending occurs due to reversible changes in turgor. (A pulvinus is the swollen base of a leaf or leaflet which contains a high proportion of thin-walled parenchyma.)

We have already mentioned one growth movement, that of phototropism, where movement of an organ occurs in response to unilateral illumination. A tropic movement of this type is thus a response to an external stimulus. In the case of phototropism the stimulus is light, but other external stimuli such as gravity, water, chemicals, heat or mechanical contact can also induce growth movements.

Where the direction of the response is related to the direction of the stimulus, we speak of a *tropic* response, or *tropism*; but in many cases the direction of movement does not bear a direct relation to that of the stimulus, and we then speak of a *nastic* response. Thus, a curvature of a shoot towards the more illuminated side is a phototropic curvature, while the opening or closing of flowers with a change of light intensity all round is a *photonastic* one. All tropic responses are induced by directional or unilateral stimuli, whereas in nastic responses the stimulus may be diffuse. Another example of a nastic growth response is seen in the opening and closing of certain flowers, such as those of crocus, in response to changes in temperature (thermonasty). When the temperature rises growth is faster on the inner side of the petals, so that the flower

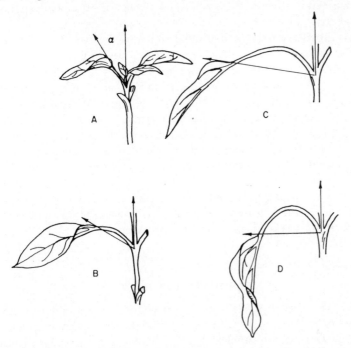

FIG. 65. Principal changes in orientation of the lamina and petiole during the growth of the leaf in *Helianthus annuus*. A. Young leaf emerging from the terminal bud. B. Leaf during the phase of rapid growth. C. Mature leaf. D. Leaf approaching senescence. α = orientation angle. (Reprinted from J. H. Palmer and I. D. J. Phillips, *Physiol. Plant*, **16**, 572–584, 1963.)

opens, whereas the reverse is true when the temperature falls. Not all nastic movements are growth movements, some being brought about by reversible turgor changes in pulvini, as in the movements of French bean (*Phaseolus vulgaris*) leaves in response to light and dark.

We shall consider phototropism and geotropism in more detail below, but a few examples of other types of response may be mentioned here. The tropic movement of a plant organ in response to an external chemical stimulus is named *chemotropism*. An example is seen in pollen tube growth down the style towards the ovules, for the directional stimulus for the elongating pollen tube is provided by certain chemicals present in the ovule and ovary wall, although their precise nature is not known. The familiar sight of tendrils of a pea plant, or other climbing plants, twining around a support is a good example of a response to contact or mechanical stimulus, and is called *haptotropism* (or *thigmotropism*). Many roots respond to differences in soil moisture content and grow towards regions of wetter soil (*hydrotropism*). There are many other examples of both tropic and nastic responses which, for reasons of space, we cannot describe further here.

A few growth movements in plants do not obviously result from any external stimulus, but appear to be spontaneous movements arising from causes within the plant itself. Examples are straight extension growth, nutations and epinastic movements. A *nutational movement* is seen in the growth of a plant stem, for it does not grow straight upwards, but performs a series of rhythmic movements which result in the shoot tip oscillating about the longitudinal axis. This particular type of nutation, called circumnutation, is very pronounced in both shoots and tendrils of climbing plants, and may confer a biological advantage in the finding of a support. An example of an *epinastic movement* is seen in the petioles of many species, which show a growth curvature, the upper side growing more rapidly than the lower, resulting in a downward movement of the lamina as the leaf grows (Fig. 65). This downward movement of leaves with increasing age is known to result from an internal stimulus, and not from an external one such as gravity.

PHOTOTROPISM

This is the growth movement of a plant organ in response to unilateral illumination and, as we saw earlier (p. 63), studies on this phenomenon led to the discovery of auxins. In general, stems and other aerial portions of plants are positively phototropic (i.e. they bend towards the light source), while roots and other underground organs are negatively phototropic (bend away from the light source). There are, however, many exceptions to these rules; for example, some tendrils and stems are negatively phototropic, and many roots non-phototropic or even positively phototropic when young, becoming negatively phototropic only later on. It is found that the response depends on both the intensity of illumination and the period of exposure, so that, within certain limits, the response is determined by the quantity (i.e. intensity × time) of light received. Although, within certain limits, the same quantity of stimulus produces the same response whether high intensity illumination is given for a short time or low intensity for a longer time, the magnitude of the response is not proportional to the quantity of stimulation. Thus, in the case of the oat coleoptile (Fig. 66) it is found that with increase in the quantity of stimulus there is an increase in the response until a maximum is reached, above which, with increasing quantity of stimulus, the response falls off until at a certain value the initial "positive" curvature is reversed and a "negative" curvature (i.e. away from the illuminated side) occurs. With still greater quantities of stimulus the curvature again becomes positive.

In studies of the role of auxin in phototropism most use has been made of the coleoptile of oat (*Avena*), wheat (*Triticum*), maize (*Zea mays*) and barley (*Hordeum*) seedlings. The reasons for this are the convenience which such material offers in experimentation, and the great sensitivity of coleoptiles to unilateral illumination.

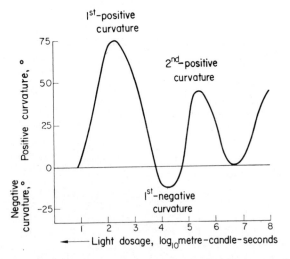

FIG. 66. The relationship between phototropic curvature in the Avena coleoptile and dosage of unilateral light. (From H. G. du Buy and E. Nuernbergh, *Ergeb. Biol.* **10**, 207–322, 1934.)

We have seen earlier (Fig. 37) that various early experiments led to the conclusion that there is a transmission of auxin from the coleoptile tip to the basal tissues of the shaded side. Over the years, several theories of phototropism involving auxin as a correlation factor have been proposed. These are:

1. The *Cholodny–Went theory*, which was put forward independently by Cholodny and Went in the 1920's. This theory suggests that a phototropic stimulus induces a *lateral translocation of auxin* across the photosensitive region of a coleoptile, leading to a higher auxin concentration in the darker half which consequently elongates more rapidly than the illuminated side—i.e. a positive phototropic curvature ensues.

2. A theory invoking *photodestruction of auxin* in the illuminated tissues (see

also p. 68), which produces a differential in auxin concentration between illuminated and darkened regions.

3. The suggestion that the rate of auxin *synthesis* is lower in the illuminated than in the darker parts of the coleoptile tip.

Apart from these three theories, there is a fourth concept which has been held in the past, that of a *light-growth reaction* not necessarily involving auxin. This suggests that when light impinges on a plant cell it affects its growth rate, either stimulating or suppressing it. However, it is known that suppression of cell extension growth by light (as would be necessary for a positive phototropic curvature to develop) is achieved by the effect of the red wavelengths (p. 179),

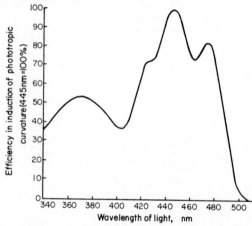

FIG. 67. Action spectrum for first-positive phototropic curvature in the *Avena* coleoptile. Maximum curvature occurs with unilateral light of the blue wavelengths. (From K. V. Thimann and G. M. Curry, *Comparative Biochem.* **1**, 243–306 Academic Press, 1960.)

whereas phototropism is induced by blue light wavelengths (Fig. 67). Therefore, although a red light-growth reaction is important in many morphogenetic phenomena (e.g. overall stem extension growth and leaf expansion), it does not appear to play a primary role in phototropism. At the present time it is generally considered that blue light in some manner induces differential auxin concentrations, which in turn causes differential growth in a phototropically responding organ. The immediate problem is, however, to decide in what way a change in auxin distribution comes about.

The original evidence in favour of the Cholodny–Went theory was based upon work by Went in 1928, who made measurements of the quantities of auxin which could be collected in agar blocks placed below the illuminated and shaded sides of cut coleoptile tips (Fig. 68). It was found that unilateral light increased the proportion of total auxin in the agar blocks positioned below the shaded side. Went found that 57 per cent of the total auxin that had been' obtained from the non-illuminated tip was collected from the shaded half of a unilaterally illuminated tip, and only 27 per cent from the illuminated half (Fig. 68)—i.e. revealing a 16 per cent overall fall in total auxin yield following illumination, the remaining auxin being preferentially distributed to the darker region. The 16 per cent fall in auxin resulting from exposure to light was regarded as insignificant by Went, who concluded that a unilateral light stimulus induces lateral migration of newly synthesized auxin towards the shaded side.

In his experiments, Went used a unilateral light dosage which would be expected to induce a "System I" (or "first positive") curvature, but a number of other workers have repeated Went's experiments using light dosages which

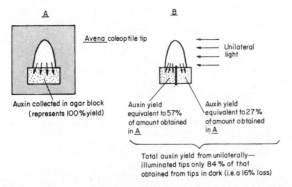

Fig. 68. Went's original experiment demonstrating that an oat coleoptile tip, when exposed to unilateral light, transmits more auxin to an agar block below the shaded side than to the block below the illuminated side (B). The total auxin yield from illuminated tips was 16 per cent less than that from tips maintained in darkness (A). This apparent loss was probably not significant (see text). (From F. W. Went, *Rec. Trav. Bot. Neerl.* **25**, 1–116, 1928.)

would induce System II ("first negative") or System III ("second positive") curvatures (Fig. 66). In all these cases an auxin differential was obtained, with the greatest amount of auxin emerging from the basal cut surface of a coleoptile tip corresponding to the most rapidly elongating side of the responding coleoptile.

Although Went regarded the 16 per cent loss in total auxin which occurred on phototropic stimulation as being insignificant, a number of other workers considered that photodestruction of auxin in the illuminated tissues may play an important part in phototropism. It was found by Galston and co-workers that blue light can be absorbed by natural plant pigments such as riboflavin, and the absorbed energy utilized in the photo-oxidation of IAA (see also p. 68). Similarly, work by Zenk has shown that a naturally occurring xanthophyll, violaxanthin, absorbs light of wavelength of 450 nm which can energize the destruction of the auxin, naphthalene-acetic acid. Nevertheless, Briggs and his associates, and Gillespie and Thimann have shown that light of the dosage necessary to produce phototropic curvatures does not decrease the overall auxin levels in coleoptile tips. There is little justification, therefore, for regarding photodestruction of auxin as playing any important part in phototropism in coleoptiles.

Even so, it is obviously of critical importance to know whether there was any real significance in the 16 per cent loss in auxin following light stimulation

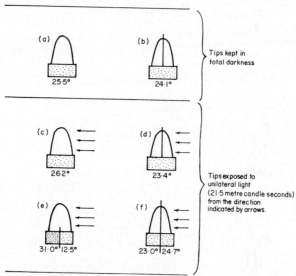

FIG. 69. Auxin diffusion into agar blocks from variously treated coleoptile tips of *Zea mays*. The figure under each agar block indicates the degrees curvature produced by that block in the Went *Avena* curvature test for auxin. The vertical line running through some of the tips and agar blocks represents an impervious glass barrier. Twice as much auxin was obtained in (e) and (f) than in (a)—(d), because each agar block had been in contact with six half-tips, the equivalent of the three whole tips placed on the agar blocks in (a) to (d) inclusive. (From W. R. Briggs, R. D. Tocher and J. F. Wilson, *Science*, **126**, 210–12, 1957.)

in Went's original experiment. Because of this, Briggs and his co-workers in California carried out a series of experiments in 1957 using corn coleoptiles (*Zea mays*), which essentially confirmed earlier work by Boysen-Jensen in that the results showed that the lateral differential in auxin concentration occurring in phototropically stimulated tips could be a consequence only of light-induced lateral transport of auxin from the light to the dark side. Briggs collected in agar blocks the auxin which diffused from coleoptile tips which had been subjected to various treatments, and then determined the quantity of auxin present in the blocks by means of the Went *Avena* curvature test. The amount of auxin produced by the tips kept in total darkness was similar to that produced by illuminated tips, whether or not the tips had been completely separated into two vertical halves by a thin piece of glass (Fig. 69a–d). This result argues against the hypothesis of photodestruction of auxin and photo-inhibition of auxin synthesis. Further, it was found that when a tip was stood on two separate agar blocks and only partially bisected at its base by a glass sheet, and was exposed to unilateral light incident at right angles to the glass plate, then significantly more auxin diffused into the agar block below the half of the tip remote from the light source (Fig. 69e). When, however, the whole of a coleoptile tip was bisected and separated by glass and exposed to the same conditions as a partially separated tip, then it was found that there was no difference in the amounts of auxin diffusing into the separate agar blocks below the "light" and

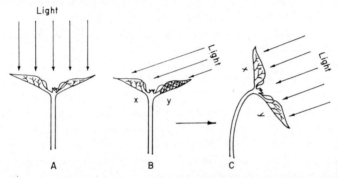

Fig. 70. Effect of light intensity on auxin production in young leaves of *Helianthus annuus*, and on the induction of a phototropic curvature in the stem. A. Both leaves illuminated equally and producing equal quantities of auxin. B. The illumination of leaf X is higher than of leaf Y, because of the difference between the angles at which the light strikes them. Hence there is greater auxin production by leaf X than by leaf Y. C. Phototropic curvature resulting from unequal auxin production by the opposite leaves. (From H. Shibaoka and T. Yamaki, *Sci. Papers Coll. Gen. Educ. Tokyo*, **9**, 105–26, 1959.)

"dark" halves (Fig. 69f). If either auxin destruction or inhibition of synthesis was responsible for the observed differential in the partially split tips one would have expected that a total glass barrier, as in Fig. 69f, would make no difference to the differential distribution of auxin observed in the partially split tips. The fact that a total glass barrier completely prevented the establishment of an unequal distribution of auxin, led Briggs to conclude that the observed differential distribution of auxin that occurs in phototropically responding coleoptile tips is a consequence of lateral movement of auxin towards the dark side, and is not a result of photo-destruction of auxin or of photo-inhibition of synthesis.

In summary, most available evidence indicates that when a coleoptile tip is illuminated from one side, there is first of all *perception* of the light-stimulus, followed by a *transverse migration of auxin* molecules within the tip towards the darker side. This in turn means that more auxin is transmitted to the region of response in the coleoptile below the darker half of the tip, resulting in greater growth of that region and consequent bending of the whole coleoptile towards the light source.

The actual mechanism whereby light is able to induce lateral movement of auxin molecules is not known. To be effective, light-energy must first of all be absorbed by the plant tissues, and there must therefore exist a *photoreceptive pigment*. Since it is blue light which induces phototropism the photoreceptor must absorb selectively light of this part of the spectrum, and reflect the other wavelengths in the visible spectrum, so that it should appear yellow. It is thought, therefore, that pigments such as riboflavin, violaxanthin, or carotenoids are the operative photoreceptors in phototropism. Recently, evidence has been presented that illumination reduces the rate of polar auxin transport in coleoptiles and this is accompanied by reduced growth. It is possible that this effect is important in phototropism.

The phototropic mechanism of leafy shoots of dicotyledonous plants may be different from that of coleoptiles. In such leafy shoots, it has been found that most of the auxin required for elongation of the stem comes from the young expanding leaves near the apex. In other words, the lamina of a young leaf exports auxin via the petiole into the stem. Consequently, any phototropic curvature of such a stem must be a result of some alteration in the distribution, or quantity, of auxin coming from the leaves. In the case of sunflower plants (*Helianthus annuus*), the leaves are arranged in pairs on opposite sides of the stem (decussate arrangement) and it has been found that each leaf of a pair will, if both are illuminated to the same degree, supply equal quantities of auxin to the stem (Fig. 70A). If, however, one leaf of a pair is more brightly illuminated than the other because of its orientation in relation to the incident light (Fig. 70B), then the leaf receiving a higher intensity of light produces a greater quantity of auxin than its partner. This perhaps results in the side of the stem beneath the brightly illuminated leaf receiving more auxin and consequently growing at a more rapid rate than the other side, causing the stem to execute a positive phototropic curvature, until the position is reached whereby both leaves receive light at equal angles of incidence (Fig. 70C).

GEOTROPISM

This term is applied to growth movements induced by a gravitational stimulus. Growth of an organ towards the centre of the earth is termed *positive geotropism*, and growth away from the centre of the earth *negative geotropism*. When the axis of an organ comes to lie at right angles to the direction of the gravitational field it is said to be *diageotropic* (e.g. rhizomes of Solomon's seal, couch grass, etc., or stolons of potato and strawberry). Where an organ becomes oriented at intermediate angles (i.e. between 0° and 90°, or between 90° and 180° from the vertical), it is said to be *plagiogeotropic* (e.g. lateral branches are very often so). Most main roots are positively geotropic. The rhizomes of many mosses are sometimes positively geotropic, but, in general, positive geotropism is not well developed in lower plants. Negative geotropism is shown by the stems of higher plants, by the sporangiophores and sporophores of many fungi, and by the foliage shoots of mosses. While many rhizomes and stolons are diageotropic, lateral stems and lateral roots of the first order and foliage leaves are commonly plagiogeotropic. Lateral shoots and lateral roots of a higher order generally possess little geotropic sensitivity and are consequently said to be *ageotropic*.

Moving a plant from its usual vertical position to a horizontal one causes gravity to act across the width of stem and root. This results in growth curvature responses, whereby the stem bends to grow upwards and the root bends to grow downwards. This can easily be demonstrated with a young seedling, such as that of *Zea mays* or mustard.

Clearly, geotropic growth curvatures result from unequal growth rates along the upper and lower sides of the responding organ. As in phototropism, it can be shown that perception of a gravitational stimulus occurs in the tip, whereas the response occurs in the elongating zone, at a short distance from the tip. Thus, if the apical few millimetres of either a coleoptile or a primary root are cut off, then the organs no longer respond to a gravitational stimulus. However, replacing the tips with a layer of gelatine between tip and stump allows a curvature to develop, indicating that, as in phototropic phenomena, perception of the stimulus takes place in the tip, following which there is evidently a transmission of chemical stimulus from the tip to the region of response.

By analogy with the mechanism of phototropism, it might be expected that the differential growth occurring on the upper and lower sides of a horizontal organ involves an unequal distribution of auxin in the region of response, and there is much evidence in support of this conclusion. Various experiments have shown that auxin moves in the direction of the gravitational field, that is downwards. This means that when a plant is placed horizontally, higher auxin concentrations are created along the lower sides of the stem and root.

The first experiment demonstrating that auxin moves laterally downwards across a horizontal organ was performed in Holland by Dolk, in 1936, although the original proposal that this took place was put forward independently by Cholodny in 1924, and by Went in 1926. Dolk placed excised coleoptile tips of *Avena sativa* and *Zea mays* horizontally, and collected the auxin which diffused from the upper and lower halves. More auxin was obtained from lower halves than from upper halves of tips of both species, while the total auxin yield of the organs was the same as that of similar but vertical tips. A number of other workers have repeated Dolk's experiments using various organs from other plant species, and found similar distributions of auxin between the top and bottom halves (Fig. 71). More recently it has been found that when radio-active IAA (IAA^{-14}C) was applied to the apical ends of horizontal sections of the coleoptiles of *Zea mays* and *Avena sativa*, and of the hypocotyls of *Helianthus annuus*, it emerged asymmetrically into agar receiver blocks at the basal ends, so that more IAA^{-14}C was found to have passed into the lower than into the upper receiver block.

Thus it has been clearly established that a lateral migration of auxin occurs under the influence of a gravitational field as well as of a light gradient. The higher concentration of auxin along the lower side of a stem or coleoptile presumably induces a greater growth in that region compared with the upper side, bringing about upward curvature. In the case of a root, it is thought that

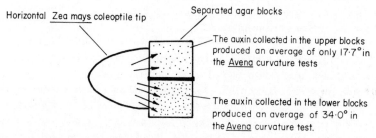

Fig. 71. Greater amounts of auxin diffuse from the lower side of a horizontally posi-tioned coleoptile tip than from the upper side. The amount of auxin present in each agar block was measured by means of the Went *Avena* curvature test (see Chapter 4). (From B. Gillespie and W. R. Briggs, *Plant Physiol.* **36**, 364–8, 1961.)

the auxin concentration on the lower side increases to become supra-optimal and consequently inhibitory to the growth of the auxin-sensitive root tissues (see Fig. 51). On the basis of this theory, the upper side of a horizontally placed root contains auxin at a more nearly optimal level and consequently grows at a more rapid rate than the inhibited lower side, resulting in a downward growth curvature of the root. Whether the production of ethylene is involved in the supra-optimal auxin inhibition of root growth is not yet clear (p. 105).

Geoperception

As is the case in studies of phototropism, no completely adequate explanation can yet be offered as to why it is that auxin migrates across a geotropically stimulated organ. There must exist some sort of *geoperceptive* mechanism, which "perceives" the direction of gravity in relation to the orientation of the organ and which is able to influence auxin to move in a particular direction. The nature of this geoperception is not yet known, though over the years a number of possibilities have been suggested.

As we have seen, removal of the coleoptile or root tip greatly reduces the geotropic sensitivity of the organ, although the region of growth curvature is several millimetres back from the tip. Hence it would seem that the primary site of perception is in the tip region. Recent experiments have shown that in maize roots removal of part of the root cap greatly reduces geotropic sensitivity whilst not affecting root elongation. This suggests that perception may occur in this part of the tip region. We have, therefore, to explain how the tip region of a coleoptile or root is able to perceive the orientation of the organ in relation to gravity.

At first sight it might be supposed that a positive geotropic growth curvature in a horizontal organ is the result of its own weight, which sets up differential mechanical stresses on its upper and lower surfaces due to a cantilever effect. However, positive geotropism is still evident in an organ whose weight is supported along its entire length, and indeed positively geotropic organs, such as roots, will curve against very strong stresses acting in the opposite direction. On this and other evidence which we need not consider here, it is now generally considered that the geoperceptive mechanism is contained within individual living cells.

Plant cells are, of course, made up of many components, and it is only by an effect upon the masses of one or more of these components that gravity can exert an influence. Consequently, it would appear that gravity influences the distribution of subcellular constituents having a density different from that of the surrounding cytoplasmic medium. For example, it is likely that oil drops would tend to *rise* in an aqueous medium such as protoplasm. On the other hand, more dense bodies included in the protoplasm, such as starch grains, mitochondria, etc., would tend to *fall* within the cell. The name *statoliths* has been given to gravity-sensitive cell inclusions such as starch grains, etc., and the term *statocyte* has been given to a cell which is gravity-sensitive and also contains statoliths. It has been suggested, therefore, that it is the pressure exerted by sedimenting statoliths, particularly starch grains which are abundant in certain cells of the root cap, upon the cytoplasm against the lower lateral wall of the cell which brings about the excitation caused by a gravitational stimulus (Fig. 72).

Certain experimental results have, however, caused doubts to be cast upon this idea; for example, some plant organs (cold-treated *Sphagnum* moss) do not contain any cells which possess starch grains, and yet these organs still exhibit geotropism. Nevertheless, in many organs it has been found that the time required for starch grain sedimentation in statocytes is closely correlated with the presentation time of the organ in which they are contained. (The *presentation time* is the time for which a stimulus of a given intensity must be applied, before a response can occur.) A convincing example of this close correlation is seen in the temperature relationships of starch grain sedimentation velocity and of presentation time in seedling stems of *Lathyrus odoratus* (Fig. 73); this correlation would be expected if the presentation time depends upon the rate of sedimentation of starch grains.

Regardless of the nature of the geoperceptive system itself, gravity-stimulated tissues must possess a mechanism which brings about the lateral migration of auxin molecules. It is possible that this mechanism is an electrical one, for it has been found that a *geoelectrical potential* is set up across both shoots and roots

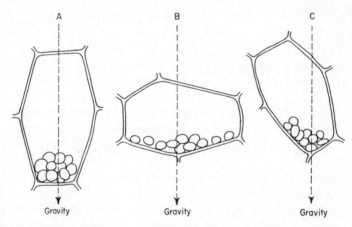

FIG. 72. Diagrammatic sections through statocytes following geotropic stimulation in three different positions. In each case the statoliths (starch grains) sedimented to the lowermost face of the cell. (From L. Hawker, *Ann. Bot.* **46,** 121, 1932.)

when they are positioned horizontally. Measurements of this electrical potential difference have shown that it amounts to from 5 to 20 millivolts, with the lower side positive with respect to the upper side. Thus, it is possible that the negatively charged ions of dissociated auxin move to the lower side under the influence of this electric field. However, recent work has shown that a lateral potential does not develop until curvature of the organ has commenced. This suggests that lateral displacement of auxin occurs *before* the geoelectrical potential is set up. Thus, it must be admitted that at present we have no idea how a displacement of starch grains in cells in the tip region could bring about a differential auxin distribution on the two sides of a horizontal organ.

Growth movements other than those of phototropism and geotropism cannot at the present time be explained in terms of the auxin relationships of the tissues, since insufficient research has been directed towards these problems.

However, some experiments have indicated that plagiogeotropic and epinastic organs possess, in addition to the geotropic system, a mechanism which causes movement of auxin preferentially to the *upper* side, thus partially counter-acting the influence of gravity and causing the organ to become oriented at an angle between vertical and horizontal.

It is not possible at the present time to evaluate possible roles of gibberellins or cytokinins in plant growth movements. Little experimental work has been conducted yet on these problems. However, it is known that gibberellin

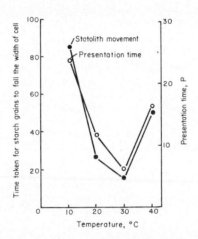

Fig. 73. Positive correlation observed between presentation time in geotropism of *Lathyrus odoratus* seedling stem, and time taken for starch grain sedimentation at different temperatures. (From L. Hawker, *Ann. Bot.* **47**, 505–15, 1933.)

treatment often results in a growth movement of the leaves of plants such that their adaxial surfaces come to lie closer to the stem. That is, gibberellins often produce hyponastic movements in leaves, whereas auxin applications normally result in epinastic movements of these organs. Also, circumnutational move-ments of the stem in *Convolvulus* and pea are enhanced by gibberellin treatment.

FURTHER READING

General

AUDUS, L. J. *Plant Growth Substances*, 2nd ed., L. Hill Ltd., London, 1959.

LAETSCH, W. M. and R. E. CLELAND (Ed.). *Papers on Plant Growth and Development*, Little, Brown & Co., Boston, 1967.

LEOPOLD, A. C. *Plant Growth and Development*, McGraw-Hill, New York, 1964.

WILKINS, M. B. (Ed.). *Physiology of Plant Growth and Development*, McGraw-Hill, London, 1969.

More Advanced Reading

AUDUS, L. J. The mechanism of the perception of gravity by plants. *Symp. Soc. Exp. Biol.* **16,** 197, 1962.

BRIGGS, W. R. The phototropic responses of higher plants. *Ann. Rev. Plant Physiol.* **14,** 311, 1963.

THIMANN, K. V. and G. M. CURRY, Phototropism and phototaxis, in *Comparative Biochemistry* (Ed. M. Florkin and H. S. Mason), Vol. 1, p. 243, Academic Press, New York, 1960.

THIMANN, K. V. Studies on the movement of auxin in tissues and its modification by gravity and light in *Regulateurs Naturels de la Croissance Végétale* (Ed. J. P. Nitsch), Edition du Centre National de la Recherche Scientifique, Paris, 1964.

WILKINS, M. B. Geotropism. *Ann. Rev. Plant Physiol.* **17,** 379, 1966.

Various articles in *Encycl. Plant Physiol.*, volume 17.

CHAPTER 8

Sterile Culture Methods in
Studies of Differentiation

HISTORICAL ASPECTS OF PLANT TISSUE CULTURE

In the earlier chapters of this book we considered various aspects of morphogenesis and we described a number of experimental approaches to the subject, including surgical techniques. No one aspect of botanical research is able to provide a unified picture of differentiation, and studies of this subject must aim at the unification of information derived from the fields of morphology, physiology, biochemistry and biophysics. This is a difficult task, for in the intact plant there are complex interactions between the various processes underlying growth and differentiation. Consequently, it is desirable to reduce the complexity of the system so that the controlling processes may be more easily identified and studied. This can be done to varying degrees, by isolating embryos and other parts of a plant in order to eliminate certain complicating correlative influences during studies of their behaviour. In general, isolated embryos and plant parts must be carefully nurtured to keep them alive and free from infection by micro-organisms. This normally entails the use of *sterile culture techniques*. Such methods often facilitate the maintenance of isolated embryos, organs and tissues for considerable periods of time. However, the culture of isolated plant tissues, organs and embryos presented many difficulties to earlier workers which were only gradually surmounted.

The smallest viable unit of a plant one can at present envisage as reproducing, growing and developing in culture is a single cell. As long ago as 1902, Haberlandt attempted to grow single higher plant cells in sterile culture, but, for various reasons which we now understand, his attempts were unsuccessful.

Following Haberlandt, other workers established methods which would allow the growth of isolated plant organs and tissues in culture. Excised roots were the first plant organs to be successfully brought into sterile culture, and work by White in the 1930's demonstrated that given appropriate nutrients,

147

such roots would grow and differentiate normal root tissues. Work by Gautheret (1939) and others established that isolated portions of storage tissue, e.g. from carrot roots, could be kept alive and grown in sterile culture. *Callus cultures* derived from such isolated tissues lend themselves to studies of the effects of nutrients, vitamins and hormones upon cell division, differentiation of vascular tissues, and the inception of organized meristematic regions within the predominantly parenchymatous tissue. By definition, a callus is a mass of proliferating tissue consisting predominantly of parenchymatous cells, but in which differentiation may occur under suitable conditions.

When a callus is grown in agitated liquid culture, cells at the surface are often broken away and float free in the medium to give a *liquid suspension culture*. Usually, such free cells do not divide when retained in a medium suitable for callus growth, for propagation of free plant cells demands a more elaborate medium than does a callus. Media have, however, been devised which are capable of supporting proliferation of free plant cells.

ORGAN CULTURE

Root Culture

It has proved possible to grow several types of plant organ in sterile culture, including roots, shoot apices, leaves, flower parts and fruits. The nutrient requirements for such organ culture vary considerably from species to species and according to the type of organ in question, but certain general requirements can be recognized. Intact higher plants are autotrophic; that is, they are able to synthesize all the organic substances required for their own life from carbon dioxide, oxygen and mineral nutrients. However, since most sterile cultures are unable to carry on photosynthesis, it is clear that they will require at least a carbon source, usually supplied in the form of a sugar such as sucrose or glucose. In addition, sterile cultures require the same mineral nutrients as the intact plant, including both macronutrients (nitrogen, phosphorus, potassium and calcium) and micronutrients (Mg, Fe, Mn, Zn, etc.).

In addition to those requirements for a carbon source and mineral nutrients, it is found that most isolated organs have also a requirement for certain special organic substances. Thus, most isolated roots grown in sterile culture require to be supplied with certain vitamins, particularly thiamin (vitamin B_1) and sometimes pyridoxin (vitamin B_6) nicotinic acid and others. It appears that in the intact plant, certain vitamins are synthesized in the leaves and that roots are dependent upon the shoots for the supply of these substances, which they are unable to make for themselves. Tomato roots require only sucrose, mineral nutrients and thiamin, and given these they will grow successfully in culture for many years.

Excised roots of some monocotyledonous plants fail to grow even when supplied with a full complement of B-vitamins and other vitamins. In some of these cases (e.g. rye), an exogenous auxin supplement to the nutrient medium allows growth to proceed. In order to maintain the culture, the excised roots must be regularly subcultured on to fresh medium, by excising a piece of root bearing a lateral, which then proceeds to grow rapidly and maintain the culture.

Excised roots of most species produce only root tissues in culture, but there are some species the cultured roots of which regenerate shoot buds as well as further roots, e.g. *Convolvulus*, dandelion (*Taraxacum officinalis*) and dock (*Rumex crispus*).

Culture of Shoot Apices and Leaves

Isolated shoot apical meristems and leaf primordia can also be grown in sterile culture. These frequently produce adventitious roots and can eventually develop into complete plants. The shoot apices and leaves of vascular cryptogams, such as ferns, are relatively more autotrophic than those of angiosperms. Thus, even a small fern apex can be grown on a medium containing only a carbohydrate source and mineral nutrients. Small angiosperm apices (less than 0·5 mm in diameter) require a general source of organic nitrogen and certain specific amino acids and vitamins, in addition to the basic medium, but larger apices will grow on a simple medium. The simpler requirements of large apices may be due to the fact that they carry larger leaf primordia, which apparently can supply some of the requirements of the apex for vitamins and other organic nutrients.

Isolated young leaves of the fern, *Osmunda cinnamomea*, and of sunflower (*Helianthus annuus*) and tobacco (*Nicotiana tabacum*), have been successfully grown on a simple medium containing only sucrose and inorganic salts (Fig. 23). Such isolated leaf primordia continue to grow and develop into normally differentiated leaves, although they are usually very much smaller than normal leaves developed on the plant (p. 33).

EMBRYO CULTURE

We have earlier (p. 25) given an account of the development of embryos as it occurs naturally within the embryo sac. Experimental approaches to the study of plant embryogenesis have in recent years made extensive use of sterile culture of aseptically isolated young embryos.

The easiest method of culturing an embryo is to allow it to develop *in situ* within the ovary or dissected out ovule. The ovule, if placed on a suitable nutrient medium, is able to support the development of the zygote to maturity. The presence of placental tissue aids the development of the ovule and embryo, and consequently it is easier to maintain ovules in sterile culture when they are left within the ovary. The physiological requirements of ovules do not appear to be species specific, since young fertilized ovules of widely different species have grown to mature seeds following transplantation on to the placenta of *Capsicum* fruits.

Although embryos can be grown from the zygote stage to maturity when they are left inside a cultured ovule, they show complex nutrient requirements when isolated from the ovule. The mature embryo is, of course, autotrophic and will grow if provided with the normal conditions necessary for germination, viz. adequate water and oxygen and a favourable temperature. However, if embryos are excised at younger stages it is found that they will not grow, even if supplied with sucrose and mineral salts.

A useful technique for the culture of young embryos was introduced in 1941 by van Overbeek, who found that they could be grown from a quite immature stage by supplementing sucrose and mineral salts with coconut milk, which is a liquid endosperm. This endosperm allows the development of the coconut palm embryo, being a complex range of substrates necessary for the growth of the embryo, and it is also very effective in supporting the growth of embryos of other species. Attempts have been made to identify the active components of coconut milk and they are now known to include sugar-alcohols, such as myo-inositol, leucoanthocyanins, cytokinins, and probably also auxins and gibberellins. By using this technique it has been possible to develop to maturity isolated embryos of certain species at an early stage of development.

As embryos develop they appear to become less heterotrophic, as has been demonstrated by observations on isolated embryos of *Capsella* and *Datura*, which showed that globular embryos did not survive at all in culture, but that early heart-shaped embryos (see Fig. 13) would develop further if supplied with a nutrient medium containing sugar, inorganic salts, vitamins and coconut milk. With slightly older heart-shaped embryos the coconut milk may be replaced by a source of reduced nitrogen such as L-glutamine, and with more highly developed embryos even the glutamine can be omitted. Thus, there appears to be a progressive increase in the synthetic abilities of the embryo during its development.

On the other hand, there is more recent evidence that the concept of a decrease in heterotrophic nutrition with increase in embryo age may not be strictly true. Even young globular embryos of *Capsella* have been cultured to maturity in a relatively simple medium containing no organic nitrogen source such as L-glutamine, nor high sucrose concentration. But, in addition to the usual inorganic salts, vitamins and 2 per cent sucrose, the medium did have to contain a balanced mixture of very low concentrations (about 10^{-7} M) of an auxin (e.g. IAA), a cytokinin (e.g. kinetin) and adenine sulphate. It appears that, in the case of *Capsella* at least, it is the balance of growth hormones in the immediate environment rather than substrate level metabolites, which plays a determining role in embryogenesis.

TISSUE CULTURE

In contrast to the techniques of organ culture, tissue culture involves the sterile culture of an isolated homogeneous mass of cells. All plant organs consist at the time of their excision of a number of different tissue types, and therefore represent more complex systems than do isolated individual tissues. In studies of the physiology and biochemistry of morphogenesis it is desirable to work with as simple a system as possible, and clearly culture of isolated tissues would appear to represent a simplification of experimental material in comparison with organs in culture. Fortunately, tissues from many sources can be maintained in culture for an indefinite period, and afford enormous but as yet largely untapped, possibilities for research in physiology and biochemistry. Such cultures have already proved very valuable for certain biochemical studies. For example, cultures of sycamore (*Acer pseudoplatanus*) cambial tissue have been extensively used for studies on cell wall metabolism, and we shall see below several examples of the value of tissue culture for studies on differentiation.

When small pieces of root phloem parenchyma of wild carrot (*Daucus carota*), or pith parenchyma of tobacco (*Nicotiana tabacum*) stem, or even chlorophyll-containing palisade cells from leaves of *Arachis hypogea* and *Crepis capillaris* are placed on a suitable medium, they can not only be kept alive but can be induced to grow. That is, mature parenchymatous or mesophyll cells, which, if left undisturbed in the plant body, would undergo no further cell division, can be made to divide mitotically, giving rise to an undifferentiated *callus*. An extreme example of retention of the capacity for cell division in mature plant cells was provided by cultivation of a callus from medullary ray tissue excised from a region adjacent to the pith in 50-year-old lime (*Tilia*) trees. These cells had matured a full half-century earlier, and yet their continued potential for active cell division was revealed under suitable conditions in culture.

It seems likely, therefore, that any living, nucleated, plant tissue can give rise to a proliferating undifferentiated callus when excised and placed on a suitable culture medium. However, great difficulty is often experienced in establishing a callus from a previously untried source of tissue, for there is apparently considerable diversity in nutritional requirements of tissues from different species, or even from different locations within one plant. In general, it has proved easier to culture tissues consisting originally of non-green parenchyma, such as phloem or pith parenchyma. The establishing of green, photosynthesizing, callus growths from chloroplast-containing leaf cells has come much later.

In addition to the usual macro- and micro-inorganic nutrients, and an organic carbon source, isolated tissues are frequently found to require (1) an organic source of reduced nitrogen, which may be supplied as amino acids or, in some species, as the amide of glutamic acid, L-glutamine; (2) vitamins, including thiamin, nicotinic acid and pyridoxine, and (3) the sugar alcohol, myo-inositol. In addition, an auxin, such as 2,4-D, and sometimes a cytokinin, are required. The fact that it is necessary to supply hormones to callus cultures, whereas they are not normally required by organ cultures, may indicate that the organized meristems of organ cultures may be centres of hormone biosynthesis, whereas the parenchymatous tissues from which callus cultures are derived do not have the capacity for hormone synthesis.

It is of interest to note that photosynthetic callus growth from palisade mesophyll cells of *Arachis hypogea* does not need an external supply of any vitamins. This can be related to what was said earlier concerning the normal production of vitamins in the leaves of plants, and their supply to other regions such as the roots.

Repeated subculturing of some tissue cultures, such as those of carrot, grape, *Scorzonera*, tobacco, and other plants, leads to a spontaneous and irreversible change in that they acquire the capacity to synthesize excess quantities of auxin. Thus, a tissue which when first brought into culture requires an exogenous supply of auxin, later on, after subculture, becomes autotrophic for auxin. Such long-established callus cultures are then said to be *habituated*, or *anergized*, and closely resemble tumorous as opposed to normal plant tissues. For example, plants infected with the crown-gall bacterium (*Agrobacterium tumefaciens*, synonym *Phytomonas tumefaciens*) exhibit tumorous (callus-like) growths at the points of infection. By appropriate heat treatment, the bacteria can be killed and removal of a portion of one of these treated tumours into sterile culture leads to the production of a massive undifferentiated callus, which is completely self-sufficient in auxin. Thus, infected or habituated cells have undergone a permanent change in that they are able to synthesize substances which they were unable to produce before. This capacity is transferred from one cell generation to the next (see p. 283).

FREE CELL CULTURE

The principal problem encountered in attempts to make isolated single cells divide in culture is that free cells are "leaky". For a number of reasons, particularly their large surface area exposed to the medium, in contrast to callus cells surrounded by like cells, individual free plant cells tend to lose substances required for cell division to the medium. Consequently, the nutritional requirements of free plant cells may be more complex than those for callus cultures of the same species, since it is necessary to supply the substances which tend to be lost by the cells into the medium. For some types of cell culture it has been possible to determine the precise nutrient requirements so that they can be grown on a defined medium, which usually included the various constituents already listed for callus cultures (p. 152). In other cases, it has not yet been possible to grow free cell cultures on a defined medium and it is necessary to add coconut milk, which must, therefore, include certain, as yet unknown, special nutrient factors.

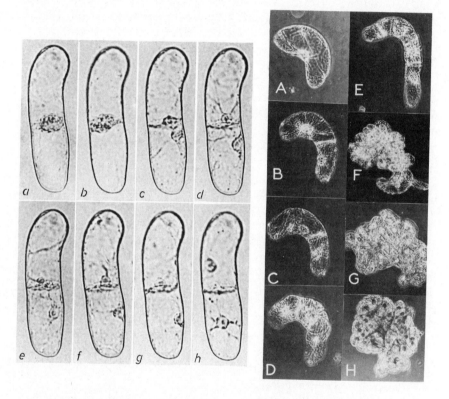

Fig. 74. *Left*: Cell division in a single isolated cell from *Phaseolus vulgaris*. Note the large vacuole, prominent cytoplasmic strands and large nucleus with nucleolus. Time lapse between pictures a, b and c, 30 minutes; between c, d, e, f, g and h, 60 minutes. (Reprinted by permission of the Rockefeller University Press from the *Journal of General Physiology*, **43**, 843, 1959–60. Print supplied by Dr. Ludwig Bergmann.)

Right: Development of a cluster of cells from a single cell of tobacco stem–pith isolated in sterile culture. A. Single cell one day after placing in culture medium. B–H. Stages in the formation of a mass of cells, from the single cell in A. (From W. Vasil and A. C. Hildebrandt, *Science*, **150**, 889–92, 1965. Print supplied by Dr. A. C. Hildebrandt.)

Although all the nutritional problems involved in growing colonies of cells from different sources have not yet been solved, we are able now to see the general picture. What is striking is the very close similarity between nutritional requirements for free cell culture and for embryogenesis in isolated young embryos (p. 150). In the intact plant these requirements are met by surrounding tissues—particularly, in the case of embryos, by the endosperm.

Suspensions of free cells growing in culture look very much alike from whatever species they originated. The principal characteristics of a free cell are: (i) numerous and large vacuoles, even in cells capable of division, (ii) prominent cytoplasmic strands which show active streaming movements, (iii) a large nucleus with nucleolus (Fig. 74A). In addition, a variety of cell types coexist in a given cell suspension, not all of which remain as single cells. Some cells divide and give rise to clusters of smaller, more dense cells. Others increase in size and divide with the formation of internal cross-walls to produce either a filament of cells or in some cases, a new free cell by a process analogous to "budding" of yeast cells in culture (Fig. 74B). Thus, free cells of higher plants do not have the morphology of cells in the tissue from which they were derived. Furthermore, they evidently have a different pattern of metabolism, for they usually do not contain typical storage products.

Ideally, the initial inoculum of isolated cells in a culture vessel will consist exclusively of individual free cells. No multicellular clusters should be present. The ideal can be attained—for example by filtering the suspension through a sterile gauze having a pore diameter too small to allow the passage of cell clusters. By such means a very dilute suspension of free cells (5 or less cells per ml of nutrient medium) can be set up. Under suitable conditions this suspension of cells will multiply so that in 2 or 3 weeks' time the cell density will have risen to approximately 100,000 per ml. Microscopic examination of the cell population at this time reveals that not all the cells are now free—i.e. cell clusters of various sizes and shapes are usually present in addition to single cells. Various studies have shown that formation of a multicellular cluster in a suspension culture of plant cells takes place by repeated division of one cell, the daughter cells of which do not separate. Separate free plant cells do not aggregate into clusters in the way that some cultured animal cells do.

REGENERATION STUDIES WITH STERILE CULTURE

Plants show remarkable capacities for regenerating whole organisms from isolated pieces of shoot, root, or leaf, and even from relatively unorganized tissue such as callus. Some more general aspects of regeneration will be discussed later, but we shall first describe experiments on regeneration which have been carried out with callus and free cell cultures.

In callus cultures cell division occurs randomly in all directions and gives rise to an unorganized mass of tissue; thus, there are no clearly defined axes of polarity in a callus. By contrast, in a shoot or root meristem we have a highly organized tissue structure, in which, as we have seen, quite well-defined patterns of division can be recognized. It has been found that under certain cultural conditions shoot and root meristems can be formed within a callus, so that whole new plants may be regenerated.

Root and Bud Regeneration in Callus Cultures

We have already made mention in Chapter 4 (p. 75) of Skoog's studies on the interacting influences of auxins and cytokinins on the growth of tobacco pith-derived tissue cultures. Skoog observed that cytokinin and auxin interacted to initiate cell division, but he also found that these same two types of growth hormone could interact to initiate organized meristems. Thus, it was discovered that if the *proportions* of auxin and cytokinin were varied, then the pattern of

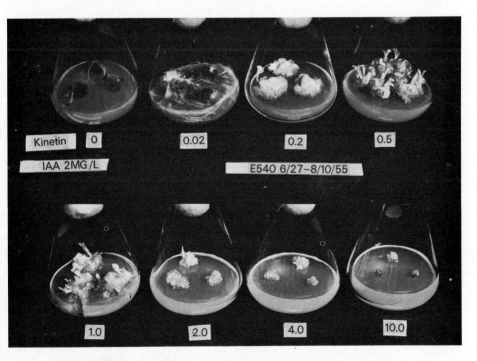

FIG. 75. Effect of a range of concentrations (0–10 mg/l) of kinetin on growth and organ formation in tobacco-pith derived callus cultured on nutrient agar containing, in all flasks, 2 mg/l indole-3-acetic acid. All cultures 44 days old. (Reprinted from *Symp. Soc. Exp. Biol.* **11**, 1957. Print kindly supplied by Professor F. Skoog.)

meristem formation was altered. When the proportion of auxin to cytokinin was relatively high, there was differentiation of some callus cells into root primordia. A higher concentration of cytokinin relative to auxin caused cells to differentiate into shoot apical meristems. Subsequent growth of the root and shoot primordia led to the callus cultures having the appearance shown in Fig. 75. Thus, small changes in the auxin–cytokinin ratio could (a) initiate meristems and (b) channel differentiation of these into either shoot or root apical meristems.

Control of root or shoot-bud formation in callus cultures by variations in auxin–cytokinin balance has now been demonstrated for tissues of several origins. The interacting effects of the hormones in this phenomenon can be modified by other factors, such as sugar and phosphate levels, sources of nitrogen, and other constituents of the medium such as purines. There is, however, no doubt that auxin and cytokinin may regulate not only the initiation of organized meristematic centres in the callus, but also the type of meristem formed. Nevertheless, stimuli other than auxin and cytokinin are involved in apical meristem initiation in callus cultures. For example, initiation of lateral roots in pea stem segments is inhibited by red light. Thus a phytochrome-based mechanism (p. 177) may be involved in the initiation of root apical meristems.

Embryo Formation in Callus and Free Cell Cultures

The formation of normal plant embryos in tissue cultures was first observed by Reinert in 1959. Following a succession of deliberate changes in the composition of the nutrient media, callus cultures of carrot root phloem parenchyma gave rise to perfectly normal embryos which would, if transferred to a suitable medium, develop into whole carrot plants (embryos formed from cells of various origins in sterile culture are often referred to as *adventive embryos*). The changes in the nutrient media required to bring about adventive embryo formation principally involved alterations in the balance of auxin and cytokinin.

Since this first report, a number of other workers have obtained adventive embryos in callus cultures from a number of plant sources. It is, however, with cultures of tissue derived from the wild strain of *Daucus carota* that greatest success has been achieved in embryo formation. Steward has demonstrated that scrapings of tissue from immature ovular embryos of carrot will produce a friable callus in culture, which when transferred to an agitated liquid medium containing coconut milk breaks up to form a suspension of cells. Plating out the cell suspension onto agar containing the same nutrient medium resulted in the formation of literally thousands of carrot embryos, each of which could develop into a whole plant. It was noticed that in some cases the embryos formed on the agar medium were derived from what were originally single cells—the first unequivocal demonstration of totipotency in plant cells. Embryos formed in this way pass through the typical stages of embryogenesis seen in normal ovular embryos, and may germinate and grow into adult carrot plants (Fig. 76).

Callus cultures derived from the root, hypocotyl, stem and petiole of wild carrot have all been shown to be capable of giving rise to adventive embryos, but those derived originally from embryo tissues show a greater tendency to form adventive embryos than cultures from older tissues, which suggests that previously matured tissues, even when "dedifferentiated" by being brought back to a meristematic state in culture, revert less readily to the embryonic condition.

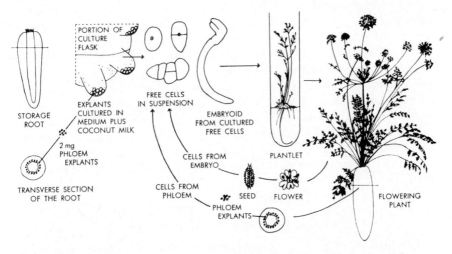

FIG. 76. Diagram to show the sequence of growth from tissue explants isolated from the secondary phloem of the root to proliferating cultures which give rise, in liquid medium, to freely suspended cells which develop into adventive embryos, plantlets, and mature carrot plants, repeating the cycle indefinitely. Free cells which originate from the embryo may also behave in a similar way. (From F. C. Steward *et al.*, *Brookhaven Symp. Biol.* **16**, 73, 1963.)

To date, most experiments have involved culturing tissues on media containing the liquid endosperm of coconut. This might lead to a suspicion that coconut milk contains special unknown nutritive or hormonal "embryo-inducing" factors. Such a view does not, however, appear to be justified, for Halperin was able in 1964 to cause the production of adventive embryos in callus derived from carrot root growing on a complex, but fully defined, agar medium.

Regeneration of Vascular Tissue

Higher plants show considerable capacity for regenerating vascular tissue following wounding or grafting. Thus, we saw earlier (p. 58), that when a vascular bundle of *Coleus* is severed, a new cambium is formed between the cut ends of the bundle and gives rise to new xylem and phloem, which re-establishes the continuity of the bundle; we also saw that IAA from the upper leaves appears to be necessary for vascular regeneration in *Coleus*. Regeneration of vascular tissue can also occur in isolated plant parts growing in culture; for example, work by Camus in 1949 with chicory (*Cichorium intybus*) callus cultures showed that when a chicory shoot bud was grafted onto the top of the callus, then within a few days some stimulus had passed out from the implanted bud into the underlying undifferentiated callus cells, inducing the formation of vascular tissue which become connected to that in the bud (Fig. 25B). The stimulus which came from the grafted bud and which caused the initiation of vascular tissues, was shown to be chemical in experiments in which the stimulus was found to pass through a layer of cellophane placed between the bud and the underlying callus cells. Experiments by Wetmore and Sorokin on lilac (*Syringa vulgaris*) callus confirmed the earlier results of Camus, and further revealed that auxins such as IAA and naphthalene acetic acid (NAA) could produce the same effects in the callus as an implanted bud. That is, vascular tissue could be induced to differentiate in the callus either by grafting on a lilac shoot bud or by applying an auxin. The inference was, therefore, that the grafted bud produced auxin, and it was this which induced differentiation of vascular tissue in the callus mass.

REGENERATION IN SHOOT AND ROOT CUTTINGS

Although not normally performed in sterile culture, one of the most obvious examples of regeneration is seen in the common horticultural practice of vegetative propagation of plants by taking shoot or root cuttings, and allowing them to develop adventitious roots and/or buds. In shoot cuttings, a callus is frequently formed at the base of the cutting, as a result of divisions originating in the cambium, and from such a callus root primordia arise. However, adventitious roots may also be formed in normal tissues of the stem—usually in the pericycle, but in some species in the cambial zone. In root cuttings, both roots and buds commonly arise from callus formed from parenchyma in the younger phloem.

The ease with which roots can be formed on shoots varies enormously; cuttings from plants such as bean will produce roots if simply left with their lower ends immersed in water, whereas those from other species will do so only rarely, even under what appear to be the most favourable conditions. In those species which do produce roots, it is generally true that a piece of stem which possesses a bud or leaves will form roots at its base, but that a disbudded and defoliated stem piece produces roots much less readily or not at all. This suggests that a substance is formed in the buds and leaves which moves downward and stimulates root formation at the base of the stem. The existence of such a root-initiating substance in young leaves has been proved by the demonstration that extracts of young leaves do stimulate the rooting response of stem cuttings. Further, it has been found that the application of auxins to cuttings has a similar effect to that of extracts of young leaves (Fig. 64), suggesting that the stimulatory effect of buds and leaves upon rooting is probably due to the production of auxin by these organs. The effect of auxin in rooting of cuttings is to increase the rate of formation and absolute number of adventitious root initials. This is, therefore, another example of cell division and differentiation being activated by auxin.

The formation of roots in stem cuttings normally occurs at the *basal end* of the stem. This is true even if the cutting is inverted, so that the morphological lower end is uppermost (p. 51). The stem therefore shows polarity in the initiation of roots. The fact that auxins are known to stimulate the formation of root initials and also that auxins move in a basipetal manner (p. 96), lead one to believe that the polarity shown in root formation is a consequence of the movement of auxin to the morphologically lower tissues, where its arrival triggers off the processes of root initiation. In fact, if an auxin is applied to the apical end of a stem cutting, then callus formation and subsequent root formation is stimulated at the base of the cutting. If applied basally, then roots are again stimulated there.

If a stem section is dipped in a solution of cytokinin it may react by producing many buds at the morphologically upper end of the stem, but few or no roots, the opposite effect to that elicited by dipping in an auxin solution. Nevertheless, as we have seen (p. 155), the stimulatory effect of auxins upon root formation may not be revealed unless the responding tissues also contained an appropriate concentration of cytokinins, since root formation involves active cell division. Root cuttings behave similarly to stem cuttings with regard to polarity of root and shoot bud initiation, and the effects of auxins and cytokinins (p. 52.) The fact that buds and roots are formed at opposite ends of isolated segments of stem or root appears to be the result of movement of auxin and cytokinin in opposite directions, a preponderance of one or the other accumulating at either end, causing either buds or roots to be initiated. Indeed, it has been found that if cuttings of chicory (*Cichorium intybus*) roots are placed under moist conditions, which favour regeneration, then certain changes occur in the distribution of the endogenous hormones within the cuttings, so that high auxin concentrations are found at the basal ends, and high cytokinin levels at the apical ends. These changes occur *before* there is any observable regeneration of buds and roots and hence they may play an important role in the pattern of regeneration. Certainly these observations are consistent with the findings that bud and root regeneration in callus cultures are associated with high cytokinin and high auxin levels, respectively (p. 155).

At the present time we know little of the translocation patterns of cytokinins in plants. Kinetin itself is apparently not translocated readily, for it remains at, or very close to, the place to which it is applied on a plant, Naturally occurring cytokinins may well behave differently though, as there is some evidence that cytokinins are synthesized in roots and translocated up into the shoot system (p. 265). Certainly it is known that some synthetic cytokinins are translocated quite readily in plant tissues.

Auxins are not the only factors concerned in root formation. A supply of sugar is necessary, as well as other nutrients. The stimulating effect of leaves on initiation in stem cuttings may be due in part to their production of nutrients, and perhaps also to other hormonal substances more specific in promoting root formation in conjunction with auxin.

Lateral Root Initiation

In contrast to the fairly ready regeneration of roots in shoot cuttings, excised roots of most species growing in sterile culture normally form only further root tissues, including the initiation of lateral roots, and only relatively rarely are shoot buds initiated. Auxin is in some way involved in the formation of lateral roots as well as of adventitious roots. Immersion of the main root of a dicotyle-donous seedling in a solution of an auxin results in a reduction of main root extension but a stimulation of lateral root initiation. The subsequent growth of the newly produced lateral roots is also inhibited by the auxin solution. Thus auxins, at other than very low concentrations, stimulate the formation of roots but inhibit their subsequent elongation. The result is that a root immersed in a solution of auxin becomes stunted, and possesses rows of newly emerged but suppressed lateral roots.

GENERAL ASPECTS OF REGENERATION

Sinnott has defined regeneration as "the tendency shown by a developing organism to restore any part of it which has been removed or physiologically isolated and thus to produce a complete whole". This broad definition includes a wide variety of phenomena, but we can distinguish a number of general aspects of regeneration. Firstly, we have seen that we can apply the term to the initiation of shoot and root meristems in a disorganized mass of callus, which may be growing in sterile culture or may form at the surface of a cutting in response to wounding. This is a remarkable phenomenon, even though it may be so familiar that we come to accept it as commonplace. We have no concep-tion as to the nature of the factors operating whereby in a mass of disorganized callus a high degree of organization emerges, but we have already suggested

(p. 42) that the apical meristem of the shoot or root is a stable configuration which, as it were, "crystallizes" out under certain conditions.

It is important to realize that in regeneration we see in operation the processes which determine normal development. When the normal course of development is disturbed by wounding or in other ways, compensating events occur which tend to restore the normal situation. Thus, it would seem that the normal form of the plant represents an equilibrium state, and that when this equilibrium is disturbed, built-in control mechanisms operate to restore the equilibrium. This phenomenon is well illustrated in the regenerative properties of shoot and root meristems. We have already seen that if a shoot apex of *Lupinus* is divided by two vertical cuts at right angles, then each segment of the original apex is able to regenerate into a normal apex (p. 42). Similar experiments have been successfully carried out with root apices. Further examples are seen in the regeneration of vascular tissue (p. 58) and in the formation of a phellogen when the surface of a stem is cut or damaged.

So far we have discussed the problems presented by the spontaneous development of organized meristems within unorganized meristematic tissue. A further problem concerns the resumption of cell division in previously differentiated, non-dividing cells, which follows wounding. In some cases regeneration of root primordia takes place in callus tissue which has developed at the cut basal surface of a shoot cutting, while in other cases the root primordia may be formed by the resumption of cell division in stem tissues, such as the pericycle. In either case, however, it is clear that the isolation of a piece of stem or other organ results in renewed cell division, and the question arises as to what causes this cell division. There is some evidence that the wounding of plant tissues results in the release of "wound hormones", which stimulate cell division, and one such hormone, "traumatic acid" ($HOOC.CH{=}CH(CH_2)_8.COOH$) has indeed been isolated. Whether such substances are involved in all cases of cell division following wounding is not clear. In any case, however, it is clear that certain differentiated cells of the stem or root become "dedifferentiated" when they resume meristematic activity.

The phenomenon of regeneration provides strong evidence that the process of differentiation in many types of plant cell does not involve any loss in their genetic potentialities, so that they remain "totipotent".

Although the totipotent behaviour of individual cells of a number of plant species has been demonstrated experimentally (Fig. 76), it is nevertheless wise to be cautious in assuming that all living, nucleated, plant cells are totipotent. Until regeneration of whole plants has been seen to occur from isolated cells of all types known to occur in the plant body (e.g. parenchyma, palisade and spongy mesophyll cells, companion cells), the case cannot be regarded as proven. Even so, as we have mentioned earlier (p. 157), Steward has regenerated whole carrot plants from adventive embryos formed from cells derived from the root, hypocotyl, stem, petiole, and embryo of wild carrot. Cells from the lamina of green leaves do appear to be more recalcitrant in demonstrating their presumed totipotentiality. A number of workers in the field of plant cell and tissue culture, particularly Steward, nevertheless consider that any free cell from a higher plant will, if provided with the right stimuli (nutritional and hormonal), regenerate a whole new plant taking one of the alternative routes toward organization described above.

Steward has argued that one of the requirements for the production of embryos from free cell cultures is that a cell must be set free from association with its neighbours and thus be able to grow independently. When cells occur in a group, the development of each individual cell may be restricted. Thus, cells growing *en masse* in a callus do not form embryos, but will do so if they are separated in a free cell culture. Conversely, the activities of callus cells may become further restricted if there is differentiation; the cells of undifferentiated callus are usually capable of unlimited division, but if a bud is regenerated, then the cells which become part of the leaf primordia are subject to considerable restraints with respect to the planes of their divisions, and they are no longer capable of unlimited division so long as they remain part of the leaf. We do not know how a cell becomes restricted when it is a part of a tissue system, but possibly some control over each cell is exerted by its neighbours through the plasmodesmata, which connect the protoplasts of adjacent cells.

The discovery that the regeneration of buds and roots by callus tissues can be regulated by the relative concentrations of auxin and cytokinin in the culture medium has led to the suggestion that these hormones play an important role in "organ formation". However, it should be noted that the primary effect of the hormones is upon the initiation of shoot and root apical meristems, and we do not find callus cultures giving rise directly to organs of determinate growth, such as stems and leaves, although these may be formed as a result of the subsequent growth of the meristems. Unless we are prepared to call a shoot meristem an "organ", which seems inappropriate, it is not strictly accurate to say that cytokinins promote organ formation, although it is perhaps more justifiable to say this of the initiation of root primordia in response to auxin.

Nevertheless, the initiation of two kinds of apical meristem in response to different hormone levels is a highly interesting effect and raises the question as to whether differences in the levels of endogenous auxins and cytokinins in different parts of the plant may play a role in normal development. The observation that there is an increase in the levels of endogenous cytokinins in the apical ends, and of auxins at the basal ends of cuttings of chicory root (p.159) and that these changes precede the initiation of adventitious buds and roots, suggests that these hormones are important in natural regeneration. We shall return to a discussion of the possible role of growth hormones in morphogenesis in Chapter 13.

FURTHER READING

General

LAETSCH, W. M. and CLELAND (Ed.). *Papers on Plant Growth and Development*, Little, Brown & Co., Boston, 1967.
STEWARD, F. C. *Growth and Organization in Plants*, Addison-Wesley, Reading, Mass., 1968.
STEWARD, F. C. The control of growth in plant cells. *Sci. Amer.* **209**, 104, 1963.

More Advanced Reading

SKOOG F. and W. MILLER. Chemical regulation of growth and organ formation in plant tissues cultured *in vitro*. *Symp. Soc. Exp. Biol.* **11**, 118, 1957.
STEWARD, F. C. and H. Y. M. RAM. Determining factors in cell growth. *Adv. in Morphogenesis*, **1**, 189, 1961.
STEWARD, F. C., A. E. KENT and M. O. MAPES. The culture of free plant cells and its significance for embryology and morphogenesis, in *Current Topics in Developmental Biology* (Ed. A. Monroy and A. A. Moscona), Vol. I, Academic Press, New York, 1966.
TORREY, J. G. The initiation of organized development in plants. *Adv. in Morphogenesis*, **5**, 39, 1966.

CHAPTER 9

The Physiology of Flowering—
I. Photoperiodism

INTRODUCTION

In the "typical" life history of a herbaceous flowering plant, there is usually an initial phase of vegetative growth, which sooner or later is followed by the reproductive phase. However, there is a good deal of variation with respect to the distinctness of these two phases; in some plants there is a fairly sharp transition from vegetative growth to reproduction, as in wheat or sunflower, while in others vegetative growth and flowering occur concurrently, as in runner beans (*Phaseolus multiflorus*) or tomato (*Lycopersicum esculentum*). In the first type of plant, growth is usually *determinate*, the axis being terminated by an inflorescence, whereas in the second group the flowers are axillary and borne on lateral shoots, while the main axis continues vegetative growth. In plants of both groups, however, there is nearly always a certain minimum period of purely vegetative growth—only very exceptionally can flowers be formed immediately following germination, e.g. in *Chenopodium rubrum*, under short-day conditions. The duration of this vegetative phase is very variable. Usually there is a certain period of growth during which a succession of new leaves is formed at the shoot apex. In some perennial species, however, the number of leaves is already predetermined in the stage of dormancy and the vegetative phase involves only the expansion of leaf primordia already laid down in the previous year, as in many bulbous plants and woody species.

Factors determining the onset of flowering

What causes the transition from the vegetative to the reproductive phase? Is it due to some internal control mechanism, determined by the genetical make-up of the species, or is it dependent upon a change in external conditions? Now, plants differ very greatly in their sensitivity to external conditions, the

164

development of some species being relatively insensitive, so that provided that the environmental conditions are not so unfavourable that growth is completely prevented, they will ultimately flower under a wide range of conditions. In other species, however, the initiation of flowering is very sensitive to external conditions and will not occur under certain conditions, e.g. of temperature or daylength, even though these may be quite favourable for growth; that is to say, the requirements for flowering are not necessarily the same as for growth.

Since our knowledge of the physiology of flowering is much more complete for species which are sensitive to environmental conditions, we shall first deal with this group, and then describe briefly what is known regarding the control of flowering in the "insensitive" group.

Quantitative measurement of flowering responses

The difference between the vegetative and the flowering condition is a qualitative one. It is important, however, that in studying the physiology of flowering, we should have an exact measure of the flowering responses to various treatments. Various indices of flowering have been used, such as (1) the percentage of flowering plants in the total group receiving a particular treatment; (2) the total number of flowers or total number of flowering nodes; (3) the time to the first appearance of flowers (the shorter the time the greater the flowering response); (4) the number of leaves formed before flower initiation; (5) the use of a scale of "scores" depending upon the stage of development reached by the flowers. This latter method is used where the flowers formed are still microscopic and incompletely developed. An arbitrary scale of "scores" is assigned to various stages of development (Fig. 77) and the total "score" of a given batch of plants is determined.

PHOTOPERIODISM

The fact that seasonal changes in daylength conditions profoundly affect the life cycle of many plants was first clearly demonstrated by Garner and Allard, two American plant breeders, in 1920. Garner and Allard were originally concerned with the peculiar seasonal flowering behaviour of certain varieties of tobacco (*Nicotiana tabacum*) and soybeans (*Glycine max*). A newly developed variety of tobacco, Maryland Mammoth, was found to grow vigorously throughout the summer, but did not flower. When grown in a greenhouse during the winter, this same variety flowered and fruited abundantly. With certain varieties of soybeans, plantings at successive intervals during the spring and summer all tended to flower at the same date in the late summer, the vegetative period being progressively shortened the later the date of sowing. Garner and Allard attempted to regulate the flowering of the tobacco and soybeans by varying the temperature, nutrition and soil moisture, but none of these factors was found to affect the date of flowering very markedly. They then investigated the effect of shortening the daily light period by a few hours by placing the plants in a dark chamber, and found that under the shortened daylight period the plants quickly initiated flowers. After this exciting discovery,

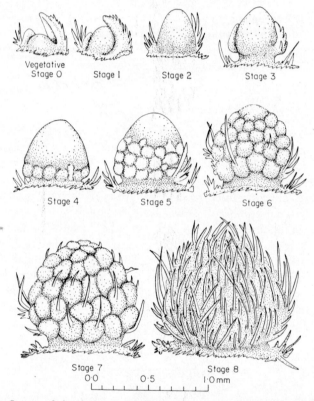

Fig. 77. Stages of development of the staminate inflorescence primordium of *Xanthium*. (From F. B. Salisbury, *Plant Physiol.* **30**, 327, 1955.)

they proceeded to test the effect of various daylength conditions on a wide variety of plant species. Variation of daylength was achieved in two ways: (1) during the summer months by shortening the natural daylength, and (2) in winter by extending the natural daylength by artificial illumination.

It was found that for many plant species the daylength (i.e. the lengths of daily light and dark periods) is a very important factor in growth and development, particularly in the control of flowering, and the phenomenon is known as *photoperiodism.* On the other hand, in other species, flowering was not materially affected by length of day. The group in which daylength has a marked effect may be separated into two subdivisions, viz. (1) those in which flowering is readily induced by exposure to short days, and known as *"short-day"* plants (SDP), and (2) a second group in which flowering is favoured by long days, and which are known as *"long-day"* plants (LDP). Species in which daylength does not markedly affect flowering are known as *indeterminate* or *"day-neutral"* plants.

Some species remain permanently vegetative if kept under unfavourable daylength conditions, and hence may be called obligate photoperiodic plants.

FIG. 78. Plants of *Kalanchoë blossfeldiana* (left) and of *Rudbeckia bicolor* (right), flowering in response to short days and long days respectively.

They include both SDP, such as *Xanthium pennsylvanicum* and LDP such as *Hyoscyamus niger*. Other species may show hastened flowering under SD or LD, but they will ultimately flower even under unfavourable daylength conditions, and hence show a *quantitative* photoperiodic response. Such species include the SDP *Salvia splendens*, rice (*Oryza sativa*) and cotton (*Gossypium hirsutum*) and the LDP wheat (*Triticum*) and flax (*Linum usitatissimum*). Obligate photoperiodic plants show a well-marked *critical daylength*, below or above which flowering will not occur in LDP and SDP, respectively, whereas facultative photoperiodic species show only a graded response to daylength, with no sharp cut-off point. Examples of SDP and LDP are shown in Fig. 78, and are listed in Table 1.

TABLE 1. SOME EXAMPLES OF SHORT-DAY AND LONG-DAY PLANTS

SHORT-DAY PLANTS

A. *Species with an Absolute or Qualitative Short-day Requirement*
Amaranthus caudatus (Love-lies-bleeding)   Ipomoea hederacea (Morning glory)
Chenopodium album (Pigweed)   Kalanchoë blossfeldiana
Chrysantheum morifolium   Lemna perpusilla (Duckweed)
Coffea arabica (Coffee)   Nicotiana tabacum (Tobacco, var.
                                              Maryland Mammoth)
Euphorbia pulcherrima (Poinsettia)   Perilla ocymoides
Fragaria (Strawberry)   Pharbitis nil (Japanese morning glory)
Glycine max (Soybean)   Xanthium pennsylvanicum (Cockleburr)

B. *Species with Quantitative Short-day Requirement*
Cannabis sativa (Hemp)   Oryza sativa (Rice)
Cosmos bipinnatus (Cosmos)   Saccharum officinarum (Sugar cane)
Gossypium hirsutum (Cotton)   Salvia splendens

LONG-DAY PLANTS

A. *Species with an Absolute or Qualitative Long-day Requirement*
Alopecurus pratensis (Foxtail grass)   Melilotus alba (Sweet clover)
Anagallis arvensis (Pimpernel)   Mentha piperita (Peppermint)
Anethum graveolens (Dill)   Phleum pratensis (Timothy grass)
Avena sativa (Oat)   Raphanus sativus (Radish)
Dianthus superbus (Carnation)   Rudbeckia bicolor (Coneflower)
Festuca elatior (Fescue grass)   Sedum spectabile (Sedum)
Hyoscyamus niger (Henbane)   Spinacia oleracea (Spinach)
Lolium temulentum (Rye-grass)   Trifolium spp. (Clover)

B. *Species with a Quantitative Long-day Requirement*
Antirrhinum majus (Snapdragon)   Petunia hybrida (Petunia)
Beta vulgaris (Garden beet)   Pisum sativum (Garden pea)
Brassica rapa (Turnip)   Poa pratensis (Kentucky blue-grass)
Hordeum vulgare (Spring barley)   Secale cereale (Spring rye)
Lactuca sativa (Lettuce)   Triticum aestivum (Spring wheat)
Oenothera spp. (Evening primrose)

The short-day group includes many plants which are indigenous to regions of low latitude, north or south of the equator, such as rice, sugar cane, hemp, millet and maize, where the daylength never exceeds more than 14 hours at any season of the year. SDP of temperate regions, where the days are long in the summer, usually initiate flowers only in the late summer as the days shorten, e.g. the cultivated "Michaelmas daisies" (*Aster* spp.) and Chrysanthemums. The typical LDP are native to the temperate regions, and flower under the naturally long days of summer. They include many of the grasses and cereals, and other common cultivated plants such as spinach (*Spinacia oleracea*), lettuce (*Lactuca sativa*), beet (*Beta vulgaris*), flax (*Linum usitatissimum*) and clover (*Trifolium* spp.), as well as many wild species. In addition to the two main groups of SDP and LDP, there is a smaller number of species with dual daylength requirements. Thus, certain species require to be exposed first to LD and then to SD for flower initiation to occur and hence are called "long-short-day" plants (LSDP); examples of this type of response are provided by *Bryophyllum crenatum* and *Cestrum nocturnum*. Other species, such as *Scabiosa succisa*, *Campanula medium* and *Trifolium repens*, require to be exposed first to SD and then to LD, and hence are called "short-long-day" plants (SLDP).

Where a species has a wide distribution, so that there is a considerable difference in latitude between its northern and southern limits, it is found that it is differentiated into a number of races or ecotypes, differing in their daylength responses, e.g. golden rod (*Solidago sempervirens*) which has a wide distribution along the western coast of North America, and perennial ryegrass (*Lolium temulentum*) which has a wide distribution in Europe and North Africa. These different forms within a given species usually show a closely graded series, from typical SDP at one end, to LD-tolerant types at the other, or from typical LDP to SD-tolerant types.

Other Responses Affected by Daylength

In addition to the onset of flowering, daylength may affect certain other purely vegetative processes in the plant. Thus, it is frequently found that the length of the internode may be much reduced under SD as compared with LD. This effect is seen at its extreme form in certain long-day species, which assume a "rosette" habit under SD, e.g. henbane (*Hyoscyamus niger*). Runner formation in strawberry (*Fragaria*) plants occurs only under LD. The strawberry has a rosette habit which is not affected by daylength, but the axillaries have extended internodes (thus forming runners) under LD.

Tuber formation is also markedly affected by daylength, and is favoured by SD, as in the Jerusalem artichoke (*Helianthus tuberosus*) and in many wild species of potato (e.g. *Solanum andigena*). (The cultivated European varieties of potato are not very sensitive to daylength, and they can form tubers even under LD.) The onion (*Allium cepa*), on the other hand, requires LD for bulb-formation.

Daylength has a marked effect on the growth and leaf-fall of many wooded plants, as will be described below.

Sensitivity to Daylength Conditions

Many plants are extremely sensitive to changes in daylength as short as 15–20 minutes, so that the flowering time of some, such as certain varieties of rice (*Oryza sativa*), may be profoundly affected by even the relatively small seasonal changes in daylength found in the tropics. Similar sensitivity is shown by cocklebur (*Xanthium pennsylvanicum*) which only flowers under daylengths of 15¾ hours or less. Evidently the mechanism whereby such plants detect changes in daylength is very sensitive.

Variation in sensitivity to daylength is shown in the number of photo-periodic cycles required to induce flowering. Thus, plants of *Xanthium* require exposure to only *one* SD cycle for flowering, and once induced to flower in this way they will continue to produce flowers for a further 12 months. The LD grass *Lolium italicum* will also flower in response to one LD cycle. However, these are extreme cases, and the majority of photoperiodic plants require more than one cycle. Moreover, even in *Xanthium*, the rate of floral development is more rapid after exposure to 2 or 3 SDs. In soybeans, the number of nodes at which flowers are found increases linearly with the number of SD cycles, up to at least 7. In many species, the daylength requirements vary considerably with the age of the plant, and frequently the minimum number of inductive cycles is found to decrease with age. Moreover, some SDP, such as soybeans, which will flower only under SDs when they are young, ultimately become capable of flowering even under LD.

LIGHT AND DARK PROCESSES IN SHORT-DAY PLANTS

A considerable amount of work has been carried out to learn more about the light and dark responses of SDP species, and our knowledge of this type of plant is considerably greater than for LDP.

It is important to ascertain, first, which part of the plant is concerned with the "detection" of daylength conditions. Clearly, in the transition from the vegetative to the flowering condition, the response is at the shoot apices, but it does not follow that "perception" of daylength conditions occurs in these parts of the plant. Indeed, it was shown by the Russian worker, Chailachjan, that the responses of SDP, such as Chrysanthemum, are determined by the daylength condition to which the *leaves* are exposed and that the shoot-apices appear to be relatively insensitive to daylength conditions. He was able to show this by exposing the leaves and shoot-apical regions independently to LD or SD conditions (Fig. 79); he found that it was only when the mature leaves were maintained under SD that flowering occurred. Thus, detection of the daylength conditions is effected by the leaves, although the response occurs at the shoot-apex. Chailachjan was quick to see that these observations imply that some "signal" must be transmitted from the leaves, which causes a response in the apices. We shall discuss the possible nature of this "signal" in a later section (p. 183).

In most plants the peak of photoperiodic sensitivity in the leaf appears to be reached when it has just attained its maximum size. At this stage quite small amounts of leaf tissue are sufficient to bring about flowering. Thus, in *Xanthium* 2 cm² are sufficient to bring about flowering under SD.

Although leaf tissue is the most sensitive to daylength conditions, the stem tissues of some species also show some sensitivity, a good example being seen in *Plumbago indica*, a SDP. Indeed, isolated internode sections of this species may be induced to flower under SD conditions in sterile culture (Fig. 80), whereas under LD they remain vegetative.

The next problem which arises is whether the responses of SDP are determined by (1) the length of the daily light period, (2) the length of the dark

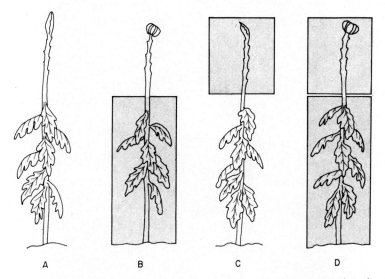

FIG. 79. Flowering occurs in Chrysanthemum when the *leaves* are exposed to short days (B and D), irrespective of whether the shoot apex is exposed to long days (B) or short days (D).

period, or (3) the *relative* lengths of the light and dark periods. Now it can easily be shown that the responses of SDP are not controlled primarily by the total duration of light which they receive each day. This is shown by the fact that soybeans will flower if exposed to three cycles consisting of 12 hours light/12 hours dark, but will not do so if exposed to 36 hours of light, followed by 36 hours of dark, or if exposed to cycles consisting of 6 hours of light alternating with 6 hours of dark (Fig. 84), although the total hours of light and dark are the same in all three regimes.

If we conduct experiments based upon the natural 24-hour cycles, it is impossible to vary the lengths of the light and dark periods independently; this can be done, however, if we use artificial sources of illumination, such as fluorescent lamps, since we can then choose any combination of light and dark periods we wish. K. C. Hamner was the first to take advantage of this fact in a series of experiments with *Xanthium* and soybean, which have now become "classical". Using soybean Hamner first investigated the effect of varying the

FIG. 80. Induction of flowering *in vitro*. Flower of *Plumbago indica* L. "Angkor" produced on a segment of stem internode (7 mm in length) excised from a vegetative plant and planted aseptically on nutrient agar. Flower formation occurs only if the culture is placed under short days (10 hours of light). Under long days (16 hours of light), vegetative buds are produced. (Experiment by J. P. and C. Nitsch. Photo. Mlle B. Norreel, kindly supplied by Dr. J. P. Nitsch.)

length of the dark period, keeping a constant duration of light period of either (a) 4 hours, or (b) 16 hours. The results (Fig. 81), showed quite clearly that soybeans will not flower until the length of the daily dark periods exceeds about 10 hours, with either 4-hour or 16-hour photoperiods. Thus, the *critical dark period* for soybeans is about 10 hours, but the maximum flowering response is reached with dark periods of 16–20 hours. In similar experiments with *Xanthium*, Hamner showed that this species has a critical dark period of $8\frac{1}{4}$ hours to flower.

Since SDP require a certain minimum period of darkness for flowering it may be asked whether they flower most readily in continuous darkness and whether they require any light. It is found, however, that although certain

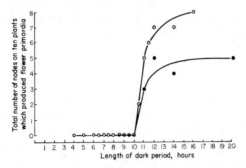

Fig. 81. Effect of various lengths of dark period, in association with constant 16-hour (–○–) or 4-hour (–●–) photoperiods, on flowering in soybean (*Glycine max*). (From K. C. Hamner, *Bot. Gaz.* **101**, 658, 1940.)

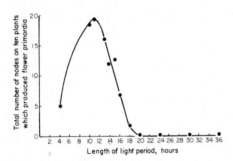

Fig. 82. Effect of varying length of photoperiod, with constant daily dark period of 16 hours, on flowering of soybeans. (From K. C. Hamner, *Bot. Gaz.* **101**, 658, 1940.)

species, especially those with a storage organ such as a tuberous rootstock, will flower in continuous darkness, other species, such as soybeans, will not do so, but require a regular alternation of light and dark.

Having investigated the dark requirements of soybeans, Hamner then proceeded to study their light requirements. Using a constant dark period of 16 hours, he varied the length of the light period and found that the flowering response increased as the length of the daily light period was increased to about 12 hours (Fig. 82), but with longer light periods fewer flowers were formed and there was no flowering when the light period reached 20 hours, even though these light periods were associated with long (16-hour) dark periods. Thus, the conditions for flowering in soybean are (1) that the daily dark period must exceed 10 hours and (2) that the length of the light must not exceed a certain duration. Of course, under natural conditions, dark periods of 10 hours or more can only be accompanied by light periods of 14 hours or less, and in nature the flowering response is likely to be controlled by the length of the dark period, rather than the duration of the light period. For this reason, it would be more appropriate to refer to short-day plants as "long-night" plants. In some SDP, e.g. *Xanthium*, long photoperiods are not inhibitory to flowering, which is determined solely by the length of the dark period.

Having thus established that flowering in short-day plants is favoured by an alternation of light and dark, the question arises as to whether the light must precede the dark or vice versa. By means of ingenious experiments with *Xanthium*, which, as we have seen, requires only one SD for flowering, Hamner was able to show that a long dark period must be *preceded* by an adequate period of high intensity light, and that a period of high intensity illumination following the dark period also promotes flowering.

The Nature of the High Intensity Light Reaction

The light requirements of SDP during the photoperiod have been investigated quantitatively by Hamner and others. In general, it appears that the light requirements are relatively high. Thus the flowering response in soybeans increases as the period of exposure to daylight is increased from 1 to 8 hours. Moreover, with photoperiods of constant duration (5 to 10 hours), the number of flowers increases steadily with increasing light intensity, at least up to 800 foot-candles.

These relatively high light requirements suggest that the process involved during the main photoperiod is photosynthesis. This conclusion is confirmed by the following facts: (1) carbon dioxide is necessary during the photoperiod, if flowering is to occur; (2) if sugar is sprayed on the leaves, then certain short-day plants can flower in complete darkness. Thus, there seems little doubt that the requirement for light in SDP is for adequate photosynthesis. This requirement might be expected, since an adequate supply of energy in the form of carbohydrates will clearly be necessary to permit the other metabolic processes, more specifically related to flowering, to occur in the leaves.

The "Dark" Reactions of Short-Day Plants

As we have seen, SDP require to be exposed to daily cycles which include a certain minimum period of darkness, the so-called "critical" dark period.

This critical dark period appears to be relatively constant and independent of the length of the photoperiods over quite a wide range of the latter. At first sight it might be thought that the dark period plays a passive role, in the sense that the effects of darkness are simply due to the absence of light effects. Thus, it might be postulated that light has an inhibitory effect upon flowering, and that the flower-promoting effects of darkness are primarily due to the absence of such inhibitory effects. However, there are several pieces of evidence which suggest that certain positive flower-promoting processes occur during the dark period. Thus, it is known that the effectiveness of the dark period increases with temperature within certain limits, suggesting the occurrence of flower-promoting processes with positive temperature coefficients during the dark period.

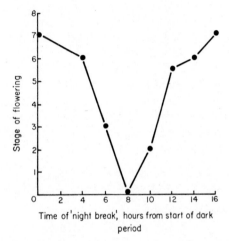

FIG. 83. Effect of night interruption, given at various times during dark period, on flowering in *Xanthium*. (From F. Salisbury and J. Bonner, *Plant Physiol.* **31**, 141, 1956.)

A very remarkable feature of the dark period is that it must be *uninterrupted* if it is to be effective in promoting flowering—interruption by only a few minutes of light during the dark period may completely nullify its effect, so that flowering is inhibited. Thus, with *Xanthium*, one minute of light at an intensity of 150 foot-candles (fc) (= 1500 lux) during a 9-hour dark period suppresses flowering. This effect has been investigated in some detail in *Xanthium*, in which it was found that a "night-break" was most effective (maximum inhibition of flowering) when given 8 hours after the beginning of the dark period, regardless of the length of the latter, at least over the range 10–20 hours (Fig. 83). It is significant that the time of maximum effect corresponds to the length of the critical dark period for this plant.

We thus have the apparently paradoxical situation that light during the main photoperiod preceding a long dark period promotes flowering, whereas light given during the dark period inhibits the flowering. We have seen that the requirement for light in the main photoperiod is probably for photosynthesis, to allow the formation of the carbohydrates and other photosynthates necessary for the metabolism of the leaf. However, it seems unlikely that the effects of a short interruption of light during a long dark period are due simply to photosynthesis, since the amount of photosynthesis occurring during 1 minute of light at 150 fc must be negligible. Thus, it would seem that we have to make a distinction between the *primary* or *high-intensity* light reaction (photosynthesis) occurring during the main photoperiod, and the *secondary* or *low-intensity* light reaction involved in the interruption of the dark period by a "night-break". The nature of this "secondary" light reaction has been studied in some detail by Borthwick and Hendricks and their co-workers at the U.S. Department of Agriculture, Beltsville, and their work has led to certain discoveries of first-rate importance, which will be described below.

Thus, we see that the flowering of SDP involves a high-intensity light reaction (photosynthesis), which must precede a long dark period exceeding a certain critical duration, and that the effect of this dark period is nullified if it is interrupted by a period of light. Before discussing further the mechanism of photoperiodic responses in SDP, the salient features of the responses of LDP will be outlined.

THE RESPONSES OF LONG-DAY PLANTS

By contrast with SDP, LDP do not require any period of darkness for flowering, and they will flower most rapidly under continuous illumination. Moreover, flowering occurs most readily if at least part of the light period is at high intensity, indicating that a certain period of photosynthesis favours flowering in LDP as in SDP. Although flowering will occur most rapidly in continuous light, it appears that long light periods are not necessary for LDP, since several long-day species will flower under regimes consisting of 6 hours of light alternating with 6 hours of darkness (Fig. 84), or even 3 hours of light alternating with 3 hours of darkness. The requirement is not that they shall be exposed to long light periods, but rather that the daily cycle shall not include long dark periods. Thus, although darkness is not essential for the flowering of LDP, nevertheless it is not without effect and long dark periods *inhibit* flowering in such plants. For example, the LDP *Hyocsyamus niger* will not flower on cycles consisting of 12 hours of light and 12 hours of dark, but it will do so under cycles of 6 hours light alternating with 6 hours dark. Thus, as in SDP, the flowering of LDP is profoundly affected by the length of the dark period, but the effects of long dark periods are opposite in SDP and LDP.

There seems little doubt that the effect of darkness on LDP is not simply due to the absence of light, but is due to processes occurring in the leaves which are actively inhibitory to flowering, since certain LDP (e.g. *Hyoscyamus niger*) will flower even under SD if they are completely defoliated. Moreover, certain LDP will flower in short days if cool temperatures are maintained during the dark period, suggesting that there are inhibitory dark processes, the rate of which is reduced at low temperature. As in SDP, the effect of a long dark period on LDP is nullified by a short interruption by light, but in the latter plants, flowering is *promoted* by a "night-break".

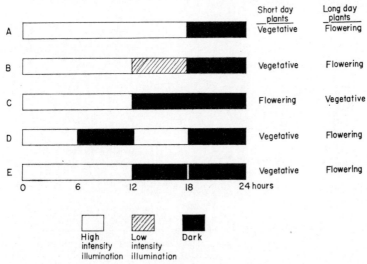

Fig. 84. Summary of responses of short-day plants and long-day plants to various photoperiodic regimes.

We thus have the somewhat paradoxical situation that in both SDP and LDP, active dark processes are involved and that the effect of darkness is nullified by short periods of illumination, but the effects on flowering are opposite in the two types of plant (Fig. 84).

THE ISOLATION AND ROLE OF PHYTOCHROME

In their classical studies extending over many years, Borthwick and Hendricks started by studying the "action spectrum" (i.e. the regions of the spectrum which have physiological activity) for the night-break effect in SDP. The problem was to produce a spectrum of sufficient purity and intensity to illuminate the leaves of soybean plants for short periods during the dark period. They overcame the technical difficulties by using a large prism and a cinema projector as a source of light. In this way they were able to produce a spectrum 2 metres in length. They used young soybean plants from which all but one of the leaves had been removed. The plants were then arranged so that the leaves occurred at various regions of the spectrum and they were irradiated for short periods. By carrying out this process with large numbers of plants they were able to determine accurately which parts of the spectrum were effective in suppressing the flowering of soybeans. They found that the most effective region was in the orange-red, from about 600 to 650 nm, with a second, much less effective band in the blue from 420 to 500 nm (Fig. 85).

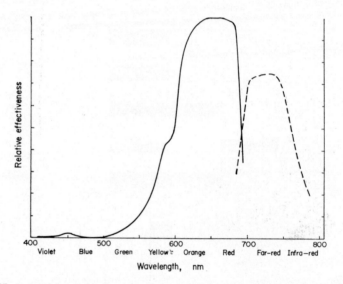

FIG. 85. Action spectra for the red and far-red physiological responses controlled by phytochrome. *Continuous line:* red effects; *broken line:* far-red effects. (Adapted from F. B. Salisbury, *Endeavour*, **24**, 74–80, 1965.)

Having determined the action spectrum for the night-break effect in SDP, they carried out similar investigations with the LDP barley and *Hyoscyamus niger*, and obtained action spectra closely similar to those of SDP, leaving little doubt that the photoreaction involved is identical in both types of plant. Moreover, they found a similar action spectrum for the removal of etiolation phenomena in peas (p. 179), and for the stimulation of germination in the light-sensitive seeds of certain varieties of lettuce (p. 240). This evidence suggested that possibly the same photoreaction may be involved in photoperiodism, etiolation and light-stimulated germination, and this conclusion was greatly reinforced by their later work.

Now, at first it was thought that other regions of the spectrum, outside the orange-red and blue, were not effective in these phenomena, but while studying the germination of lettuce seed, the Beltsville group confirmed the earlier discovery of Flint and McAlister, that far-red radiation (in the region 700–800 nm) is strongly *inhibitory* to the germination of lettuce seed. Thus red promotes, and far-red inhibits, the germination of lettuce seed. They then made the remarkable discovery that red and far-red are mutually antagonistic in their effects and that if the two types of radiation are given successively, then whether germination occurs or not depends on which was the *last* type of radiation. This suggests that far-red reverses the effect of red and vice-versa. This conclusion is supported by the fact that the effects of red and far-red may be reversed many times (Table 2).

TABLE 2. EFFECTS OF RED AND FAR-RED RADIATION ON GERMINATION OF LETTUCE SEED

Exposure to	Germination
R	+
R–FR	−
R–FR–R	+
R–FR–R–FR	−
R–FR–R–FR–R	+

It was then found that similar red/far-red interactions occur also in the photoperiodic responses of SDP, such as *Xanthium*, when the dark period is interrupted by short periods of irradiation. Thus, interruption of a long dark period by a short period of red light inhibits flowering in *Xanthium*, but if the red irradiation is followed by far-red, then the effect of the red is nullified and flowering occurs. Here, also, whether flowering occurs or not depends upon the nature of the last irradiation to which the plant was exposed. LDP also show "red/far-red reversibility", but in these plants red light during the dark period promotes flowering.

Similar phenomena are found not only in flowering, but also in leaf development. It is well known that when seedlings of many plants, such as peas or beans, are grown in the dark, the internodes become abnormally elongated, the leaves fail to develop normally, and the shoots are a pale colour, since they lack chlorophyll. These "etiolation" effects can be removed by short periods of illumination and here again it is the red region of the spectrum which is most effective. If dark-grown peas or beans are exposed to short periods of red light they form shorter internodes and more normal, expanded leaves (Fig. 86). If, however, the period of red irradiation is followed by far-red, then the effect of the red light is nullified and the plants remain etiolated. In addition to the effects already referred to, it is found that there is a considerable range of other responses which are controlled by the red and far-red radiation, including the development of chloroplasts and the formation of carotenoids and anthocyanins. Thus, these red/far-red effects appear to be involved not only in germination and flowering, but also in the normal development of leafy shoots. These formative effects of radiation are sometimes referred to as *photomorphogenesis*.

Now, the fact that germination is promoted primarily by red radiation (with a slighter effect of blue light) must indicate that lettuce seeds contain some substance which selectively absorbs in this region of the spectrum; a substance which selectively absorbs in the visible spectrum in this way is, of course, a pigment. Thus Borthwick and Hendricks were led to postulate the existence

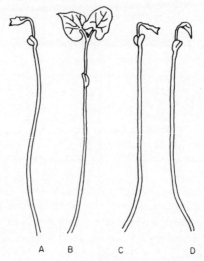

FIG. 86. Photomorphogenesis in seedlings of bean (*Phaseolus vulgaris*). Treatments: (a) grown in continuous darkness; (b) exposed to 2 minutes red light; (c) 2 minutes red and 5 minutes far-red; (d) 5 minutes far-red. (From R. J. Downs, *Plant Physiol.* **30,** 468, 1955.)

of a pigment which can be reversibly converted from one form to another. It was suggested that in lettuce seeds exposed to red light, the pigment is converted from a red-absorbing to a far-red-absorbing form and that when the latter form of the pigment is exposed to far-red radiation it is reconverted to the red-absorbing form. This hypothesis may be expressed in the form:

$$P_r \underset{\text{red}}{\overset{\text{far-red}}{\rightleftharpoons}} P_{fr}$$

where P_r and P_{fr} represent the red-absorbing and far-red absorbing forms of the pigment respectively.

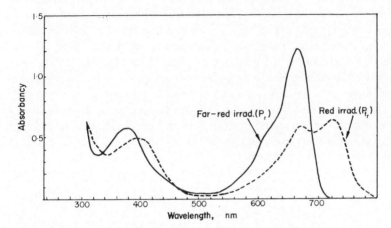

FIG. 87A. Absorption spectra of a solution of oat phytochrome following irradiation with red and far-red light, giving the P_{fr} (*broken line*) and P_r (*continuous line*) forms, respectively. (From H. W. Siegelman and W. L. Butler, *Ann. Rev. Plant Physiol.* **16**, 383, 1965.)

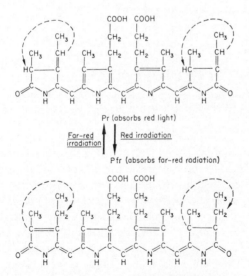

FIG. 87B. Probable structure of chromophore of phytochrome. When excited by red light, it changes from the P_r form (*top*) to the P_{fr} form (*bottom*), probably because two hydrogen atoms shift. (Adapted from S. B. Hendricks, *Scientific American*, **219**, 174, 1968.)

It should be noted that this hypothesis was put forward primarily on the basis of the observations on lettuce seed germination and at that time the existence of the pigment was only hypothetical. In 1958, however, the Beltsville workers commenced attempts to extract and isolate the hypothetical pigment. For this purpose they constructed a special spectrophotometer, by which they were able to detect any changes in absorption following exposure to red and far-red irradiation.

The changes in absorption spectrum of etiolated maize tissue were studied with this instrument. The dark grown tissue was exposed briefly to red light and the absorption spectrum of the tissue was determined. The tissue was then exposed to far-red and the absorption spectrum again determined. It was found that the absorption spectrum of the tissue had changed, so that it now

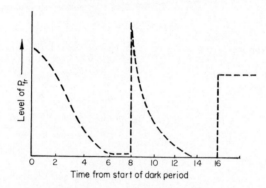

FIG. 88. Hypothetical changes in the levels of the P_{fr} form of phytochrome in leaves during a dark period which has been interrupted by a short period of exposure to red light.

showed higher absorption in the red (P_r) and lower in the far-red, as predicted by the hypothesis. Thus, to this extent the hypothesis was confirmed by direct observation. Moreover, they then found that the same properties are shown by extracts of etiolated maize shoots. Further purification of the extracts showed that the sensitive pigment is a protein. It has proved possible to effect considerable further purification and it is found that, as expected, the absorption spectrum of a solution of the pigment changes, after exposure to red and far-red radiation (Fig. 87A). This pigment they have called *phytochrome*.

Since it is a protein, it is possible that phytochrome functions as an enzyme in the plant. As with many enzymes, it consists of a protein part and a non-protein part; evidently it is the latter part (the chromophore) which is the part of molecule that undergoes reversible changes in response to red and far-red radiation. The chromophore has been shown to be a phycocyanin, a tetrapyrrole compound resembling the characteristic pigment of the blue-green algae

(Fig. 87B). It is thought that the red/far-red reversible changes involve a shift in the position of two hydrogen atoms, which results in a change in the position of double bonds. Phytochrome has now been extracted from a variety of plant species, and has been detected in both etiolated and green tissues.

Although the interconvertibility of the two forms of phytochrome in response to red and far-red light, as predicted by the hypothesis, has been directly confirmed under experimental conditions, we know much less about the changes in phytochrome occurring in the plant tissues under different daylight conditions. In etiolated tissues which have been exposed to red light the changes in P_{fr} during the ensuing period of darkness can be followed by spectrophotometer and in some species it is found that there is a corresponding increase in P_r as P_{fr} declines, suggesting that there is indeed reversion of P_{fr} to P_r, but in other species such reversion in etiolated tissues cannot be demonstrated.

Changes in phytochrome in green tissue cannot be followed with a spectrophotometer, because of interference by chlorophyll and other pigments and we can only infer the phytochrome changes from the physiological responses. It appears that during the photoperiod, under normal daylight conditions, the phytochrome contains approximately equal amounts of P_{fr} and P_r, and that during the dark period the P_{fr} declines, possibly by reversion, so that the P_r form is predominant. If the dark period is interrupted with a short period of red light, then there is rapid conversion of P_r back to P_{fr} (Fig. 88), which will then proceed to decline again during the further period of darkness.

The mode of action of phytochrome on flower induction is still not understood, but it would appear that the form P_{fr} is in some way inhibitory to the synthesis of the flower hormone in SDP, whereas in LDP P_{fr} promotes the flowering processes. We shall discuss the matter further in a later section (p. 194).

THE FLOWERING STIMULUS

We have seen that the response of the plant to daylength depends upon the conditions to which the leaves are exposed, although the response occurs at the shoot meristem. This observation immediately suggests that under favourable daylength conditions some flower promoting "stimulus" is formed in the leaves and conveyed from there to the meristems. This hypothesis was put forward by Chailachjan very shortly after his discovery that the perceptive organs in photoperiodism are the leaves, and he postulated that a flower hormone (which he called "florigen") is synthesized in the leaves under favourable daylength conditions and transmitted to the growing points. As we shall now see, there is a great deal of evidence to support this hypothesis, but nearly 30 years after Chailachjan first postulated the existence of "florigen", it still remains to be isolated and characterized chemically. A considerable number of highly interesting experiments on the transmission of the flower hormone have been carried out and these will now be described.

Hamner confirmed Chailachjan's experiments, using *Xanthium*. He showed that flowering of *Xanthium* plants under SD does not occur if all the leaves are removed, but only one-eighth of one leaf is sufficient to result in flowering

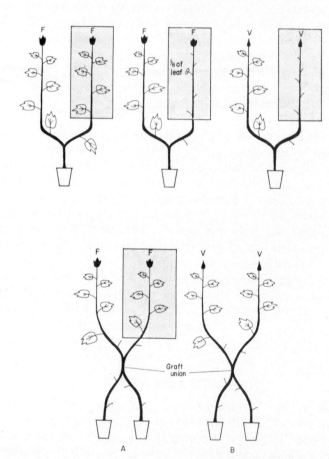

FIG. 89. *Above*: Two-branched plants of *Xanthium*, one branch of which was exposed to short days, the rest of the plant being exposed to long days. Both branches have flowered, provided that the "short-day" shoot has at least one-eighth of a leaf, but not if it is completely defoliated (*right*).

Below: A. Two *Xanthium* plants were approach grafted, and the top of one plant was exposed to short days while the other plant was maintained under long days. Both plants have flowered. B. Both plants exposed to long days; neither have flowered. (From K. C. Hamner, *Cold Spring Harbor Symp.* **10**, 49, 1942.)

(Fig. 89). In experiments with "two-shoot" plants of *Xanthium*, it was found that the stimulus arising from a single leaf is sufficient to cause flowering not only in the shoot on which it is borne, but also on the second shoot from which all leaves have been removed.

A large number of experiments have shown that the flowering-stimulus can be transmitted across a graft union. Thus, Hamner grafted together the stems of pairs of plants of *Xanthium*; when one of the plants of a pair was exposed to SD, not only did this plant flower, but also its grafted partner which had not been exposed to SD (Fig. 89). In further experiments plants of *Xanthium* were exposed to SD, their tops were then decapitated and scions from other *Xanthium* plants which had been maintained under LD were grafted on to them. In due course, it was found that the "LD" scions flowered, although they had not themselves ever been exposed to SD.

In still more striking experiments carried out with various species, including *Xanthium* and *Perilla*, single leaves were removed from "donor" plants which had been exposed to SD and grafted on to plants maintained under LD. In due course the LD "receptor" plants flowered, showing that the stimulus may be transmitted from a single leaf.

Grafting experiments with LDP have been very much fewer than with SDP, but the results have proved essentially the same.

Experiments of the type just described leave little doubt that some "stimulus", presumably a substance of a hormonal nature, is found in the leaves under favourable daylength conditions and is transmitted through the stem to the apical meristems.

Although the flower hormone has not yet been isolated from SDP, we know quite a lot about the time required for its synthesis in the leaf and about its transport in the plant. If *Xanthium* plants, each with a single leaf, are exposed to one SD cycle, and the leaves are removed at various times after the end of the long dark period, it is found that flowering does not occur unless they are allowed to remain on the plant for at least 2–4 hours following the end of the dark period, but greater flowering responses are obtained if the leaves are left for 1–2 days before removal, indicating that movement out of the leaf is a slow process. There seems no doubt that transport of the hormone occurs through living tissues, presumably the phloem, since transport is stopped by steaming a zone of stem between the leaf and the apical region, but it may be transported through other living tissues.

Attempts have been made to estimate the rate of transport of the hormone. One method was to use two-shoot plants of *Pharbitis* in which a single donor leaf on one branch was exposed to SD, and the differences in time of flower initiation on the second branch (kept under LD) were determined for varying distances between the donor leaf and the receptor bud. Methods of this type, which admittedly are very indirect, suggest that the rate of movement of the flower hormone is much slower (2–4 mm/hour) than the normal rate of trans-location of sugars in the phloem. A recent experiment with *Pharbitis* has given evidence of much more rapid transport, however (Fig. 90). Evidence was

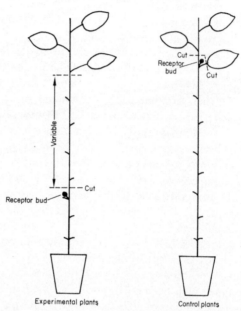

Experimental plants   Control plants

Fig. 90. *Pharbitis* plants of various heights were decapitated above the uppermost fully expanded leaf and all leaves except for the three upper ones were removed. The plants were divided into two groups. In the experimental group all axillary buds were removed except for one receptor bud at the base of the stem, and in the control group all buds were removed except for the one in the axil of the lowest donor leaf. The plants were given a single dark period which ranged in duration from 14 to 26 hours in different batches of plants, and at the end of the dark period all leaves and stems were cut off just above the receptor buds. No flower buds were formed in either group when the leaves were removed 14 hours after the start of the dark period. Flower buds were, however, initiated in many plants whose donor leaves were removed 16 hours after the start of the dark period, irrespective of the length of stem between the donor leaf and the receptor bud. The tallest plant had a stem 102 cm long between the lowest donor leaf and the receptor bud at the base, and initiated two flowers. As there was no flowering in the lowest donor leaf in control plants given a 14-hour dark period, the flowering stimulus must have moved through 102 cm of stem within the last 2 hours of the 16-hour dark period, yielding a rate of movement of more than 51 cm/hour. (From Takeba, G. and Takimoto, A., *Bot. Mag., Tokyo*, **79**, 811–14, 1966).

obtained that the flowering stimulus had moved over a distance of 102 cm in 2 hours, i.e. a rate of 51 cm/hour, which is commensurate with the rate of movement of photosynthates in the phloem.

PHOTOPERIODIC "AFTER-EFFECT" AND THE INDUCTION OF LEAVES

As we have seen, a short-day plant does not have to be maintained continuously in short days in order to flower. After a certain number of favourable photoperiodic cycles a SD plant will flower even though it may subsequently be transferred to LD. Thus, as we have seen, *Xanthium* will ultimately flower in response to a single SD cycle, although a considerable number of LD cycles may intervene between the SD and the ultimate appearance of flowers. Thus, certain species showed a marked photoperiodic "after-effect", and a plant is said to become "induced" after exposure to SD.

This after-effect of favourable photoperiodic cycles is evidently a property of the leaves, which apparently continue to produce "flower-hormone" even after they have been transferred from favourable to unfavourable daylength conditions. Thus, the inductive response of the whole plant reflects the inductive changes occurring in the leaves. This conclusion is supported by many experiments. For example, Lona subjected a single leaf of a *Perilla* plant to SD, the other leaves remaining under LD conditions. After this period of SD treatment, the leaf was allowed to remain under LD for 4 weeks, when it was removed and grafted on to a vegetative *Perilla* plant, which in due course flowered. Thus, the leaf "remembered" the previous SD treatment although 4 weeks of LD treatment intervened between the end of the last SD cycle and the grafting on to the receptor plant. What is the basis of this after-effect shown by the leaves?

Two theories have been put forward to account for the after-effect. Chailachjan postulated that a store of flower hormone is accumulated in the leaf under favourable conditions and is gradually exported from it for a long period even under unfavourable conditions. On the other hand, another Russian worker, Moshkov, postulated that the metabolism of the leaf somehow becomes *permanently* changed in response to favourable daylength conditions, so that it continues actively to *produce* flower hormone even if it is subsequently transferred to unfavourable daylengths. The available evidence seems strongly to support Moshkov's theory. Thus, it would seem unlikely that in the experiment of Lona described above sufficient flower hormone would still remain in the SD-treated leaf after 4 weeks of LD treatment to induce flowering when it was grafted on to a vegetative plant.

Even stronger evidence that leaves may become permanently changed under favourable daylength conditions is provided by certain experiments of Zeevaart. In one experiment, *Perilla* plants were first exposed to forty SD cycles and the leaves were then grafted on to vegetative plants growing under LD. Every 14 days, ten of these leaves were removed and regrafted on to a new group of vegetative plants. It was found that the leaves continued to induce flowering in each new group of vegetative stocks on to which they were grafted (Fig. 91). There was no diminution in the flowering response even with the fifth group of leaves, which had been maintained under LDs for 10 weeks. In a second experiment there was no diminution in the flower-inducing effect after seven such graftings, although more than 3 months had elapsed since the leaves had received the SD treatment. Thus, in *Perilla*, the leaves appear to become permanently induced.

It is found that the state of induction is strictly localized within the *Perilla* plant and even within the individual leaves. Thus, if single pairs of leaves of *Perilla* are exposed to a series of SD cycles and the rest of the plant is kept under LD, it is found that only those leaves which directly received SDs are capable

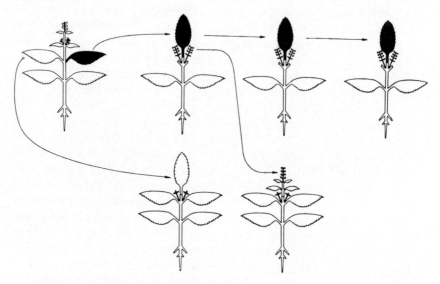

Fig. 91. Permanency of the photo-induced state in *Perilla* (see text). Black: plants exposed to short days; in outline: plants on long days. Leaves which have directly received SD treatment will induce flowering when grafted successively to negative receptor plants, whereas non-induced leaves from the original or from the receptor plants, and flowering shoots from the latter, will not do so. (From A. Lang, *Encycl. Plant Physiol.* **15**(1), 1416, 1965, after data of Zeewart.)

of inducing the plants to flower when grafted on to them (Fig. 91). In other experiments by Lona, only one half of single leaves were exposed to SD, and the other half of each leaf to LD. The leaves were then divided longitudinally and grafted separately on to vegetative receptor plants. Only those half leaves which had been directly exposed to SD were capable of inducing flowering.

The situation is different in *Xanthium*, however. If a single leaf from a flowering *Xanthium* plant (A) is grafted on to a vegetative plant (B) growing under LD, this will flower, as has already been described. If other young leaves, *which have never themselves been exposed to SD*, are taken from plant B and grafted on to further vegetative plants (C) these will also flower. Thus, the leaves of plant B have themselves become induced by the grafting of an induced leaf from plant A, although these leaves of B have never directly been exposed to SD. This effect is described as *secondary induction*, and it probably explains why *Xanthium* does not revert to the vegetative condition when it is transferred from SD to LD conditions, since all the new leaves formed will become secondarily induced. On the other hand, the new leaves formed by a *Perilla* plant which is transferred from SD to LD will not be induced and they appear to inhibit the flower promoting effects of the older leaves which became induced by the previous SD treatment (see below), and hence the plant reverts to the vegetative condition.

Although *Xanthium* and *Perilla* show marked persistence of photoperiodic induction of leaves, it is not known how general are these effects. Reversion of whole plants to the vegetative condition on exposing to a minimal number of SDs and then returning them to LD certainly occurs in soybean and a number of other species, but whether this is due to "de-induction" of the leaves exposed to SD, or to the production of new, non-induced leaves, as in *Perilla*, is not known.

As Zeevaart has pointed out, the permanent induction of leaves makes it necessary to distinguish between two distinct phenomena, viz.

(1) the *induced state* (i.e. the ability to produce the floral stimulus), gradually built up under the influence of favourable daylengths, which is irreversible and strictly localized in some plants;

(2) the *floral stimulus*, which is transmissible from induced leaves to the growing points where it exerts its morphogenetic effect.

NATURAL FLOWER-INHIBITING EFFECTS

Several lines of evidence suggest that not only flower-promoting processes, but also flower-inhibiting effects occur in photoperiodic plants. Thus, if one branch of a two-shoot plant of soybean or *Perilla* is exposed to SD and the other to LD, the latter will not flower unless its own leaves are removed. That is to say, the flowering stimulus from the "donor" branch does not produce any effect in the "receptor" branch in the presence of LD leaves. Similarly, if a scion from a vegetative plant of *Perilla* is grafted on to a stock which has previously been maintained under SD, the scion will only flower if its leaves are removed. Thus, in some SD species, LD leaves exert an inhibitory effect on flowering.

This inhibitory effect of LD leaves is only manifested if they occur between the shoot apex and the source of the flowering stimulus. Thus, if a single leaf of *Kalanchoë* is exposed to SD and the remainder of the plant is maintained under LD, the presence of LD leaves between the SD leaf and the shoot apex prevents flowering, but if all leaves are removed *above* the SD leaf and only LD leaves below are allowed to remain the plant will flower. A LD leaf is particularly inhibitory if it is immediately above the SD leaf (i.e. on the same orthostichy). LD leaves on the opposite side of the stem from the SD leaf have less inhibitory effect.

It has been suggested that these inhibitory effects of non-induced leaves can be interpreted in terms of interference with the translocation of the flower hormone, which, it is assumed, is carried with the main stream of photosynthates. The leaves which supply the greater part of the photosynthates to the shoot apical region are the uppermost mature leaves. If the latter are SD leaves, then they will supply flower hormone, as well as photosynthates, to the shoot apex, but if LD leaves are interposed between the apex and the SD leaves, then the supply of both photosynthates and flower hormone from the SD leaves to the apex will be reduced.

Other authors have argued that these inhibitory effects of leaves can only be interpreted in terms of specific flower-inhibitory substances; that is to say, they postulate that there are flower-inhibitory hormones as well as flower-promoting ones, but this conclusion is based upon indirect evidence.

Other flower-inhibiting effects are seen in experiments in which SDP such as *Kalanchoë* are exposed to SD cycles between which are intercalated one or more LD cycles. It can be shown that if the plants are exposed alternately to two SD cycles and one LD cycle, flowering is completely inhibited, and a careful analysis of the data indicates that the effect of the intercalated LD cycle is not merely "neutral" but positively inhibitory to flowering. In this type of experiment it would seem that we are dealing with inhibitory effects of LDs within the leaf itself, and not with the export of a flower-inhibitor to the shoot apex.

Flower-inhibiting effects also occur in LDP, since, as we have seen, in these plants long dark periods appear to have an inhibitory effect. It seems likely, however, that the mechanism of inhibition in this case is different from that occurring in the leaves of SDP when they are kept under LD.

It might be asked whether, if there are flower-inhibiting processes in both SDP and LDP, it is necessary also to postulate active flower-promoting processes? Might not the flowering of SDP when transferred to SD conditions be due primarily to the removal of the flower-inhibitory processes occurring under LD? If so, we may not need to postulate the existence of the elusive "flower hormone". However, most workers in this field consider that other experimental evidence points strongly to the existence of both flower-promoting and flower-inhibiting processes. Thus, the observation that a single leaf taken from a flowering plant of *Xanthium* will induce flowering in a vegetative plant maintained under LD is difficult to explain in terms simply of the removal of flower-inhibition in the receptor plant. It would seem, therefore, that the regulation of flowering in both SDP and LDP may involve the interplay of both flower-promoting and flower-inhibiting processes.

TIME MEASUREMENT IN PHOTOPERIODISM

We have seen that flowering in SDP occurs when they are kept under day-length conditions in which the length of the night exceeds the critical dark period and that some plants can detect differences in the length of the dark period of as little as 15 minutes. Thus, *Xanthium* has a critical dark-period of $8\frac{1}{4}$ hours at 25°C. A difference of only 15 minutes in the length of the dark period can determine whether or not sugar cane will flower. It is clear, therefore, that these species have rather accurate time-measuring mechanisms. Several suggestions have been made as to the nature of this mechanism.

Thus, the "clock" might be of the "hour-glass" type, in which the time taken for a particular substance to accumulate or be depleted to a certain threshold value may be the time-measuring process. Since flowering in SDP requires that the phytochrome in the leaves shall be in the P_r form, and that there is a gradual conversion of P_{fr} to P_r during the first hours of darkness (p. 183), it has been suggested that the critical dark period may represent the time taken for P_{fr} to decline to a certain level. However, it is found that the reaction $P_{fr} \longrightarrow P_r$ is effectively complete after 2–3 hours of darkness, whereas the critical dark period of *Xanthium* is $8\frac{1}{4}$ hours.

Again, if the length of the critical dark period is determined by the time taken for the conversion of P_{fr} to P_r, then irradiation with far-red light at the beginning of the dark period should greatly reduce the length of the critical dark period for flowering in SDP, by hastening the conversion of P_{fr} to P_r, but this is found not to be the case for all but one of the species so tested. On these and other grounds it seems unlikely that the rate of conversion of P_{fr} to P_r is the factor determining the length of the critical dark period of SDP. We are therefore forced to consider other hypotheses regarding the time-measuring mechanism, and among these is the "Endogenous Rhythm" hypothesis of Bünning.

ENDOGENOUS RHYTHMS IN PHOTOPERIODISM

It is many years since Bünning first drew attention to the existence of per-sistent rhythms in plants. He investigated the diurnal movement of leaves in the runner bean (*Phaseolus multiflorus*) in which the primary leaves rise during the early part of the day and later fall towards evening and reach a minimum posi-tion during the night; they then start rising again towards the morning. These movements are regulated by turgor changes in the "pulvini" of the leaves.

Now, if a bean seedling is exposed to a period of daylight and then kept in continuous darkness for several days, it will continue to show the typical diurnal movements, rising and falling on a 24-hour cycle even though it is itself not being exposed to the natural alternation of light and dark. That is to say, the bean plant shows a persistent endogenous rhythm in its leaf movements. If the position of the leaf is plotted against time then we obtain a sinusoidal curve. As the period of darkness is extended, the amplitude of the movements gradually declines to zero. The plant then needs a further exposure to light to set the rhythm in train again.

Since these early observations on leaf movements, it has been shown that many other processes in plants show a regular diurnal rhythmicity, including the following:

1. Opening and closing of flowers.
2. Root pressure.
3. Growth rate.
4. Respiration and other metabolic processes.
5. Activity of certain enzymes.
6. Mitosis and size of nucleus.
7. Discharge of fungal spores, e.g. *Pilobolus*.

Thus, there can be no doubt as to the existence of endogenous rhythms in plants. Now Bünning postulated that there is also an endogenous rhythm in photoperiodic sensitivity, and he put forward a theory of photoperiodism based upon this endogenous rhythm. He suggested that during the day SDP are in a "photophile" phase, when light is favourable for flowering, and that during the night they are in a "photophobe" phase, when darkness is favourable and light is inhibitory to flowering. Thus we have to envisage a regular rhythm in photoperiodic sensitivity, as the plants enter first the photophile and then the photophobe phase. It was postulated that SDP will flower when the daily light and dark periods correspond with the photophile and photophobe phases of the plants, i.e. when they are under SD. Under LD light will extend into the photophobe phase and will inhibit flowering. The responses of LDP have proved more difficult to account for on this theory, and these responses may be the reverse of those of SDP.

Bünning's theory has been subjected to certain criticisms and put to various experimental tests, some of which have not given the results predicted by the theory. Nevertheless, there now seems no doubt that there is an endogenous rhythm in photoperiodic sensitivity in at least some species. This rhythm has been demonstrated in various ways.

Thus, Hamner grew soybean plants on a wide range of different cycle lengths; each plant received 8 hours of light followed by a dark period the length of which varied from 8 to 62 hours. It was found that the plants flowered maximally when the total cycle length was 24 hours or a whole multiple thereof, i.e., 48 hours, 72 hours, but remained vegetative when the cycle length amounted to about 36 or 60 hours (Fig. 92). These results strongly suggest that the light and dark processes in soybean are geared to a 24-hour cycle, so that when the cycle length is 24, 48 or 72 hours in length, the endogenous "photophile" and "photophobe" phases will correspond with the environmental light and dark periods, but that when the latter are on 36 or 60 hour cycles they will be out of phase with the endogenous rhythm of the plants, which will therefore not flower. It is also found that the vegetative growth of tomato plants is best when the cycle length is 24 hours or a whole multiple thereof, but is poor on cycle lengths of 36 or 60 hours.

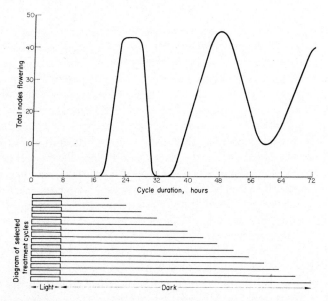

FIG. 92. Flowering responses of Biloxi soybeans to cycles by various lengths. Plants were exposed to seven cycles, each cycle consisting of 8 hours of high-intensity light (1000–1500 fc) and associated dark periods of various lengths. Total nodes flowering per 10 plants is plotted against cycle length. (After K. C. Hamner, Chapter 13 in *Environmental Control of Plant Growth*, Academic Press, New York, 1963.)

Thus, there seems little doubt that some plants do show an endogenous periodicity in photoperiodic sensitivity. It still remains an open question, however, whether such rhythms occur in all species showing photoperiodic responses, since it has proved difficult to demonstrate their existence in certain species, including *Pharbitis*. Thus, endogenous rhythms may modify the photoperiodic responses in some species but may not be a universal factor in the photoperiodism of all plant species. Moreover, the demonstration of rhythmicity by the type of experiment described above does not imply the correctness of Bünning's hypothesis regarding photophile and photophobe phases.

It is clear that the occurrence of endogenous rhythms implies the existence of some sort of "oscillator" mechanism in the plant, but the nature of this oscillator is still completely unknown. However, such an oscillator could serve as a time-measuring mechanism or physiological clock, but whether this is the actual time-measuring mechanism involved in photoperiodism must remain an open question for the present.

THE SEQUENCE OF PROCESSES LEADING TO HORMONE SYNTHESIS

We have seen that it has been possible to recognize a number of partial processes leading to the production of the flowering stimulus in SDP, and we are now in a position to summarize briefly the present state of knowledge regarding the sequence and interrelations of these processes.

Firstly, there is a requirement for a high intensity light reaction (photosynthesis) which is met during the photoperiod, and which evidently provides the energy and substrates necessary for the processes occurring during the ensuing dark period. However, there is reason to believe that phytochrome effects are also involved during the photoperiod, since the flowering of SDP, such as *Kalanchoë blossfeldiana*, when maintained in continuous darkness for long periods, is *promoted* by a short daily period of red light, suggesting that a minimal level of P_{fr} during the photoperiod is required for flowering. Normally, at the end of the photoperiod, the phytochrome will be present in approximately equal concentrations of P_r and P_{fr}. During the first few hours of the dark period, P_{fr} decays, so that P_r now predominates. Flower hormone synthesis does not commence until a certain minimum period of darkness (critical dark period) has elapsed and it then proceeds rather rapidly in the next few hours. It is apparently necessary for phytochrome to be present in the P_r form in order for flower hormone synthesis to proceed, since we know that a short interruption with red light (which converts P_r to P_{fr}) inhibits flowering. However, the commencement of hormone synthesis is apparently not directly controlled by the decay of P_{fr}, since the duration of the critical dark period is apparently considerably longer than the time required for P_{fr} decay. Hence it has been suggested that there must be some "time-measuring" mechanism, which more directly controls the commencement of hormone synthesis. We have seen that it is possible that the time-measuring mechanism involves an endogenous rhythm in certain unknown processes, and that the role of phytochrome in flowering may lie in its effects on the time-measuring processes, rather than directly on the process of flower hormone synthesis.

These ideas are summarized as follows:

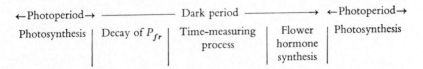

FURTHER READING

General

EVANS, L. T. (Ed.). *Induction of Flowering*, MacMillan, New York and London, 1969.

HENDRICKS, S. B. How light interacts with living matter. *Sci. Amer.* **219**, 175, 1968.

HILLMAN, W. S. *The Physiology of Flowering*, Holt, Rinehart & Winston, London and New York, 1964.

LAETSCH, W. M. and R. E. CLELAND (Ed.). *Papers on Plant Growth and Development*, Little, Brown & Co., Boston, 1967.

SALISBURY, F. *The Flowering Process*, Pergamon Press, Oxford, 1963.

WILKINS, M. B. (Ed.). *The Physiology of Plant Growth and Development*, McGraw-Hill, London, 1969.

More Advanced Reading

BORTHWICK, H. A. and S. B. HENDRICKS. Effects of radiation on growth and development. *Encycl. Plant Physiol.* **16**, 299, 1961.

FURUYA, M. Biochemistry and physiology of phytochrome, in *Progress in Phytochemistry* (Ed. L. Reinhold and Y. Liwschitz). Interscience Publishers, London and New York, 1968.

LANG, A. Physiology of flower initiation. *Encycl. Plant Physiol.* **15**(1), 1380, 1965.

LOCKHART, J. A. Mechanism of the photoperiodic process in higher plants. *Encycl. Plant Physiol.* **16**, 390, 1961.

NAYLOR, A. W. The photoperiodic control of plant behaviour. *Encycl. Plant Physiol.* **16**, 331, 1961.

SALISBURY, F. Photoperiodism and the flowering process. *Ann. Rev. Plant Physiol.* **12**, 293, 1968.

ZEEVART, J. A. D. Physiology of flowering, *Science*, **137**, 723, 1962.

CHAPTER 10

The Physiology of Flowering—
II. Temperature and Other Factors

VERNALIZATION

As we have seen, photoperiodic responses to seasonal variation in daylength conditions will account for the periodicity in flowering behaviour of many plant species, both temperate and tropical. It will be noticed, however, that among the examples of temperate species which show photoperiodic responses there were relatively few spring-flowering species, although it is a matter of everyday observation that there is a considerable number of "flowers that bloom in the spring", and many of these spring-flowering forms, such as celandine (*Ficaria verna*), primroses (*Primula vulgaris*), violets (*Viola* spp.), etc., show marked seasonal behaviour, and remain vegetative for the remainder of the year after the spring flush of flowers. It might have been expected that spring-flowering is a response to the short days of winter, but this does not appear to be the case in many species.

Daylength is not, of course, the only environmental factor showing an annual variation, and clearly temperature also shows well-marked seasonal changes, especially in temperate regions, although there is considerable variation from day to day and from year to year. We now know that seasonal variations in temperature, as well as in daylength, have a profound effect upon the flowering behaviour of many species of plant.

The clue to the importance of temperature as a regulator of flowering came first from the studies of Gassner in 1918 on the flowering of cultivated varieties of cereals. Cereals such as wheat (*Triticum*) and rye (*Secale*), can be grouped into two classes depending upon whether they are to be sown in the autumn ("winter" varieties) or in the spring ("spring" varieties). Winter wheat sown in the autumn, or spring wheat sown in the spring, both flower and mature in the following summer. If the sowing of the winter wheat is delayed until the following spring, however, it fails to ear and remains vegetative throughout the growing season. It seems unlikely that the necessity for sowing winter wheat in the autumn is simply to secure a longer growing season as such, since autumn-sown plants make relatively little growth during the winter, and winter wheat plants from spring sowings certainly seem to make adequate leaf development, yet they will not flower.

Gassner therefore investigated the effects of different temperature regimes during the germination and early growth of winter and spring rye. He sowed winter and spring rye in sand at different dates between 10th January and 3rd July and kept them at the following temperatures during germination: 1–2°C, 5–6°C, 12°C and 24°C. They were later planted out-of-doors. He found that the

Fig. 93. Effect of vernalization of flowering in "Petkus" winter rye (*Secale cereale*). *Left:* maintained for several weeks at 1°C after germination; *right:* seed unvernalized. (From O. N. Purvis, *Ann. Bot.* **48**, 919, 1934.)

temperature during germination had no influence on the subsequent flowering of spring rye, and all seedlings planted out on the same date flowered at approximately the same time, irrespective of the temperature treatment during germination. In winter rye, however, only the plants which had been germinated at 1–2°C flowered regardless of the planting date out-of-doors. Seedlings that were germinated at temperatures above 1–2°C were found to flower only if they were planted not later than March or early April, so that they would have been exposed to some actual chilling out-of-doors (in Central European climatic conditions). Gassner concluded that whereas the temperature conditions during early growth do not affect the flowering of spring rye, the flowering of winter rye depends on its passing through a cold period, either during germination or later (Fig. 93).

This work of Gassner was later followed up in the U.S.S.R. and a great deal of work was carried out there, particularly in relation to possible economic applications. The severity of the Russian winter in many regions does not permit autumn sowing of winter wheats, which, however, usually have a higher yield than spring varieties. A technique was devised by Lysenko, whereby the cold treatment required by winter wheat was applied to the seed before sowing in the spring. The method used was to allow partial soaking of the seed, so that there was sufficient imbibition of water to allow slight germination and growth of the embryo, but not sufficient for complete germination. Seed of winter wheat in this condition, which was exposed to cold treatment by burying in the snow, attained flowering and maturity in the same season if sown in the spring. The technique came to be known as *vernalization*, and this term has subsequently been extended to other treatments involving exposure to winter chillings, not only at the seed stage, but also at later stages of development of the plant.

Types of Plant showing Chilling Requirements for Flowering

We have seen that chilling of winter wheat is effective in stimulating subsequent flowering whether given during the early stages of germination or later, when considerable development of leaves has occurred. Later work has shown that many species have a chilling requirement for flowering, including winter annuals, biennials and perennial herbaceous plants. Winter annuals are species which normally germinate in the fall and flower in the early spring. They include such species as *Aira praecox, Erophila verna, Myosotis discolor* and *Veronica agrestis*.

It now appears that winter annuals and biennials are effectively monocarpic† plants which have a vernalization requirement—they remain vegetative during the first season of growth and flower in the following spring or early summer, in response to a period of chilling received during the winter. The need for biennials to receive a period of chilling before flowering can occur has been demonstrated experimentally for a number of species, including beet (*Beta vulgaris*), celery (*Apium graveolens*), cabbages (and other cultivated forms of *Brassica*), Canterbury bells (*Campanula medium*), honesty (*Lunaria biennis*), foxglove (*Digitalis purpurea*) and others. If plants of foxglove which normally behave as biennials, flowering in the second year after germination, are maintained in a warm greenhouse they may remain vegetative for several years. In regions with a mild winter climate cabbages may grow for several years in the

† See p. 255 for an explanation of this term.

open without "bolting" (i.e. flowering) in the spring, as they do in areas with cold winters. Such species have an obligate requirement for vernalization, but there are a number of other species in which flowering is hastened by chilling but will occur even in unvernalized plants; such species showing a facultative cold requirement include lettuce (*Lactuca sativa*) spinach (*Spinacia oleracea*) and late-flowering varieties of pea (*Pisum sativum*).

As well as biennial species, many perennial species show a chilling requirement and will not flower unless they are exposed each winter to cold conditions. Among common perennial plants which are known to have such a chilling requirement are primrose (*Primula vulgaris*), violets (*Viola* spp.), wallflowers (*Cheiranthus cheirii* and *C. allionii*), Brompton stocks (*Mathiola incarna*), certain varieties of garden chrysanthemum (*Chrysanthemum morifolium*), Michaelmas daisies (*Aster* spp.), Sweet William (*Dianthus*), rye-grass (*Lolium perenne*). Perennial species require revernalizing every winter.

It seems very probable that many other spring-flowering perennials will prove to have a chilling requirement when they are investigated. Bulbous spring-flowering plants such as daffodil (*Narcissus*), hyacinth, bluebell (*Endymion nonscriptus*), crocus, etc., do *not* have chilling requirements for flower initiation, the flower primordia being laid down within the bulb during the previous summer, but their growth is markedly affected by temperature conditions. For example, in tulip flower initiation is favoured by relatively high temperatures (20°C), but the optimum temperature for stem elongation and leaf growth is initially 8–9°C, rising to 13°C, 17°C and 23°C at successively later stages. Similar temperature responses are shown by hyacinth and daffodil (*Narcissus*).

In many species flower initiation does not occur during the chilling period, but only after the plants are exposed to higher temperatures following chilling. However, some plants such as Brussels sprouts, have to remain at low temperatures until flower primordia have actually been formed. It appears that all the species with a chilling requirement for flowering can be vernalized at the "plant" stage, i.e. as leafy plants, but not all species can be vernalized at the "seed" stage, as can winter cereals. Among the other species which can be vernalized at the seed stage are mustard (*Sinapis alba*) and beet (*Beta*). On the other hand, cultivated varieties of *Brassica* (cabbages, Brussels sprouts) and celery cannot be vernalized in the seed stage but the seedlings must attain a certain minimum size before they become sensitive to chilling, and thus show a "juvenile" phase. Generally, those species which can be vernalized at the seed stage are facultative cold-requiring plants, whereas those which can only be vernalized at the plant stage show an obligate chilling requirement.

Many plants with a chilling requirement resemble long-day plants, in that they have a rosette habit in the vegetative phase, and show marked internode elongation associated with flowering.

The requirement for chilling for flower initiation must not be confused with a chilling requirement for the removal of bud dormancy (see p. 230). Thus, many woody plants flower in the spring, but the flower initials are laid down within the bud during the preceding summer and they have a chilling requirement, not for flower initiation, but for the removal of bud dormancy.

Species showing both Chilling and Photoperiodic Responses

Interactions between vernalization and photoperiodic responses have been studied in a number of species. Henbane (*Hyoscyamus niger*) exists in annual and biennial forms, corresponding to the spring and winter forms of rye. The annual form does not require vernalization, but is a long-day plant, which flowers in the summer. The biennial form requires vernalization, followed by long days, for flowering to take place.

As an example of a perennial plant which shows both vernalization and photoperiodic responses, we may mention perennial rye-grass (*Lolium perenne*). In this species, flowers are initiated in response to winter chilling, but long days are required for emergence of the inflorescence, so that elongation of the flowering stem does not commence until the daylength exceeds 12 hours in March. The new tillers (lateral shoots) which emerge during the spring and summer are unvernalized and remain vegetative throughout the growing season, until the following winter. Consequently, flowering of perennial rye-grass is seasonal and restricted to the spring and early summer.

Vernalization requirements are less common among short day plants, but the garden chrysanthemum constitutes an example which has been studied intensively. Certain varieties of chrysanthemum have a vernalization requirement which must be met before they will respond to short days. After the parent plant has flowered in the autumn a number of horizontally growing rhizomes emerge from the base of the plant and grow just beneath the surface of the soil. Under outdoor conditions these new shoots will become vernalized by natural winter chilling, and in the spring they grow into normal, upright leafy shoots, which grow vegetatively under the long days of summer and initiate flowers in response to short days in the autumn. If, however, the plants are not exposed to chilling during the winter, but are grown under warm conditions in a greenhouse, the new shoots do not become vernalized, and although they grow actively throughout the summer, they are incapable of forming flowers in the short days of autumn. Thus, the chrysanthemum provides another example in which the new shoots arising each year need to be vernalized, since the vernalized condition is not transmitted from the parent shoots, as it is in rye (p. 203).

Genetical aspects of vernalization responses have been studied in several species. In rye, the difference between spring and winter forms is found to be controlled by a single major gene, the requirement for vernalization of the winter forms being recessive to the non-requirement of spring rye. The opposite situation exists in henbane, where the biennial habit (vernalization required) is dominant to the annual habit (no chilling requirement). In other species, however, the inheritance of chilling responses is more complex and in rye-grass (*Lolium perenne*) several genes appear to be involved.

Physiological Aspects of Vernalization

Intensive studies on the physiological changes underlying vernalization have been carried out on relatively few species and our knowledge of the subject is based largely on the work of Gregory and Purvis on winter rye, of Melchers and Lang on henbane (*Hyoscyamus niger*) and of Wellensiek on several other species.

The work of Gregory and Purvis has established a number of important characteristics of the processes occurring during vernalization of rye. Firstly, they showed that the changes occur in the embryo itself and not in the endosperm, as had been suggested. This was done by removing the embryos from the grain and cultivating them on a sterile medium containing sugar; such embryos were given a period of cold treatment and showed the typical hastened flowering responses given by vernalized grain, when planted. It was even possible to vernalize isolated shoot apices, which had been removed from the embryos and cultivated under sterile conditions. Such apices developed roots and regenerated into seedlings which ultimately flowered in response to the earlier chilling treatment. Moreover, it was shown that vernalization may even be effected whilst the young embryos are still developing in the ear of the mother plant. This was done by enclosing the developing ears in vaccum flasks packed with ice, or by removing developing ears and keeping them in a refrigerator until mature. In this way it was shown that vernalization is effective in embryos, even if commenced only 5 days after fertilization.

It has been shown that in older plants it is the shoot apical region which must be chilled. This was shown for celery, beet and chrysanthemum by placing a cooling coil around the shoot apical region. Thus, whereas in photoperiodism it is the *leaves* which are the sensitive organs to daylength, in vernalization the shoot apex itself is sensitive to the chilling temperatures. Wellensiek has shown that in *Lunaria* young leaves are also capable of being vernalized, but older leaves, which have ceased growth, do not respond, unless they show some cell division at the base of the petiole. Wellensiek maintains that only tissues which have dividing cells are capable of being vernalized.

For the majority of species, the most effective temperatures are just above freezing, viz. 1–2°C, but temperatures ranging from −1°C to 9°C are almost equally effective. Hence freezing of the cells is not necessary to bring about the changes occurring during vernalization, suggesting that physiological, rather than purely physical processes are involved. This conclusion is confirmed by the observation that cold treatment of rye grain is ineffective under anaerobic conditions, indicating that probably aerobic respiration is essential. By cultivating excised embryos on media with and without sugar it was shown that a supply of carbohydrate is necessary during cold treatment. Thus, although the metabolism of most plants is considerably retarded at cool temperatures, there seems little doubt that vernalization involves active physiological processes, but the nature of these processes is still quite unknown.

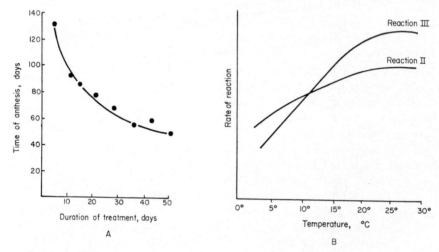

Fig. 94A. Effect of the duration of chilling of seed, on subsequent flowering behaviour of Petkus winter rye. Curve indicates time to anthesis from planting, for various periods of vernalization. (From O. N. Purvis and F. G. Gregory, *Ann. Bot.* **1**, N.S. 569, 1937.)

   B. Hypothetical scheme, illustrating two reactions with different temperature coefficients (see text).

It is found that the degree of hastening of flowering in rye varies with the duration of exposure to cold, the longer the cold treatment the shorter the period from sowing to flowering, up to a certain limit, beyond which further cold treatment has no further hastening effect (Fig. 94A). Quite short exposures to cold, such as 7–11 days, have a noticeable vernalizing effect, and this effect increases progressively with the duration of treatment.

   The vernalization process can be reversed by exposing the grains to relatively high temperatures (25–40°C) for periods of up to 4 days. Seeds treated in this way show a reduced flowering response, and are said to be "devernalized". As the period of chilling increases, it becomes increasingly difficult to reverse the effect, and when vernalization is complete, high temperature is ineffective. Devernalized grains can again be vernalized by a further period of chilling.

Once the rye plant has become vernalized it appears that the condition is transmitted to all new tissues formed subsequently, so that all new laterals are also vernalized. Indeed, if the main shoot apex is removed, so that the laterals are stimulated and then these are decapitated to stimulate secondary laterals and so on, it is found that even 4th order laterals are still fully vernalized although the apices of these laterals were not present at the time of the chilling treatment. Thus, the vernalized condition is transmitted from a parent cell to its daughter cells in cell division and it does not appear to be "diluted" in the process.

The Nature of the Changes occurring during Vernalization

One of the striking features of vernalization is that, at first sight, it appears to involve processes which go on more rapidly at lower than at higher temperatures. This effect is most unusual for chemical processes and yet we must assume that the changes occurring during vernalization are essentially enzyme-controlled reactions showing the usual characteristics of such reactions. How then are we to explain the apparent "negative temperature coefficient" of vernalization? One very simple hypothesis which has been put forward postulates the occurrence of two separate processes competing for a common substrate, each having positive (though different) temperature coefficients:

$$\text{Precursor (A)} \xrightarrow{\text{ I }} \underset{\substack{\downarrow \text{ III} \\ \text{D}}}{\underset{\text{product (B)}}{\text{Intermediate}}} \xrightarrow{\text{ II }} \text{End product (C)}$$

In this scheme reactions II and III compete for the common "Intermediate product" B. Suppose reaction III has a higher temperature coefficient than reactions I and II (Fig. 94B). This means that high temperatures will favour reaction III and more of B will be diverted into this reaction, so that little C will be formed. When, however, the temperature is markedly reduced, this will reduce the rate of reaction III more than that of II (since by definition reaction III shows greater response to changes in temperature). Consequently, reaction II will be favoured at the reduced temperature and C will accumulate. C will, therefore, be formed at lower temperatures but not at higher temperatures. Thus, the overall production of C will appear to have a "negative temperature coefficient", although each of the three involved reactions has a positive temperature coefficient. There is no direct evidence to support this hypothesis, but it is of value as indicating how an overall process may proceed more rapidly at lower temperatures, without contravening the normal laws of chemical reactions.

We have already seen that vernalization appears to involve relatively stable changes, so that once the fully vernalized state has been attained by meristematic tissue, it appears to be transmitted by cell-lineage without "dilution". This conclusion in turn may imply that the vernalized state is transmitted through some self-replicating cytoplasmic organelle, but it is equally possible that certain genes become activated during vernalization and that once this has occurred, this change is transmitted to the daughter nuclei during division.

The Flowering Stimulus in Vernalization

Although so much study has been devoted to vernalization, our understanding of the nature of the physiological and biochemical processes involved is still very fragmentary. It is not yet clear whether a specific transmissible flower hormone is formed as a result of vernalization, although there is evidence that this may be the case, at least in some species. We have seen that in photoperiodism some of the strongest evidence for the existence of a flower hormone is provided by grafting experiments, and somewhat similar results have been obtained with certain species showing a vernalization requirement.

A notable example of flower induction by grafting in a vernalized plant is provided by henbane. If a leaf from a vernalized plant of the biennial variety of henbane is grafted on to an unvernalized stock of the same variety, the latter is induced to flower without chilling. A similar response can be obtained by grafting on to the unvernalized biennial variety a leaf from any of the following: (1) the annual variety of henbane (a LDP, with no vernalization requirement); (2) *Petunia hybrida*, an annual LDP; (3) tobacco, day-neutral variety; (4) tobacco, variety "Maryland Mammoth" (SDP), under either SD or LD. Thus, transmission of a flowering stimulus can take place between cold-requiring and non-cold-requiring plants, even of different genera. Similar results have been obtained in the biennial species beet (*Beta vulgaris*), cabbage (*Brassica oleracea*), carrot (*Daucus carota*) and *Lunaria biennis*. On the basis of these results, Melchers and Lang suggested that a transmissible flowering stimulus, which they called *vernalin*, is formed as a result of chilling in biennial plants.

On the other hand, in some species, including *Chrysanthemum*, it has not proved possible to obtain transmission of a flowering stimulus from vernalized to non-vernalized plants by grafting. Moreover, if the tip region of the plant was given localized cold-treatment and hence flowered, the other buds not directly chilled remained vegetative. Similarly, when the tip of an unvernalized radish plant was grafted on to a vernalized one, flowering did not occur. These latter results are consistent with the conclusion that the *vernalized condition* (as opposed to a flowering stimulus) is only transmitted through cell division, i.e. by "cell lineage". It would seem therefore, that we must make a distinction between the vernalized ("thermo-induced") state and the formation of a flowering hormone, just as we saw that in photoperiodism we have to distinguish between the induced state of a leaf, and the transmissible stimulus formed in it.

The question then remains as to whether there is a specific flowering stimulus, vernalin, formed in plants with a chilling requirement. In most of the successful grafting experiments, referred to above, the donors were from species which have a requirement for both chilling and LDs or for LDs alone, and the experiments were usually carried out under LD conditions. Moreover, it is generally found necessary that the donor shoots should have leaves. Thus, it seems possible that "vernalin" is identical with the flower hormone produced in the leaves of LDP. Melchers and Lang have argued against this last conclusion, on the ground that leaves of Maryland Mammoth tobacco will induce flowering in stocks of biennial henbane, both under SD and under LD, although "Maryland Mammoth" itself will not flower under the latter conditions. However, the existence of vernalin must remain conjectural on the present evidence.

THE NATURE OF THE FLOWERING STIMULUS

Attempts to Isolate the Flowering Stimulus

As we have seen, the evidence from the various types of grafting experiment described in Chapter 9 strongly suggests that a flowering stimulus arises in the leaves of both SDP and LDP under favourable daylength conditions and is transported to the meristems where it causes a vegetative apex to change to the flowering condition. Moreover, it would seem that the same type of flowering stimulus is produced in both SDP and LDP, as the following experiment suggests. *Kalanchoë blossfeldiana* is a SDP, whereas the species of *Sedum ellacombianum* and *S. spectabile* of the same family (Crassulaceae) are LDP. If vegetative shoots of *Sedum* are taken from plants growing under SD and grafted on to *Kalanchoë* under SD, not only do the stocks of the latter flower, but also the *Sedum*, which will not itself flower under SD (Fig. 95). Thus, a stimulus is produced in the SDP *Kalanchoë* which will cause flowering in the LDP *Sedum*. Conversely, if vegetative shoots of *Kalanchoë* are taken from plants growing under LD and grafted on to *Sedum* plants under these conditions, again both stock and scion will flower. Thus, the LDP *Sedum* produces a flowering stimulus under LD which is capable of causing the SDP *Kalanchoë* to flower. These experiments suggest that the flowering stimulus is identical in both SDP and LDP.

A range of experimental results of the type described are consistent with the hypothesis that a specific flower hormone arises in the leaves under inductive daylength conditions, and is transmissible by grafting. As we have seen, grafting

experiments with species having a chilling requirement have also given evidence
of a transmissible flowering stimulus.

Although the circumstantial evidence for the existence of flower hormones
is very strong, repeated attempts to extract a specific flower hormone from

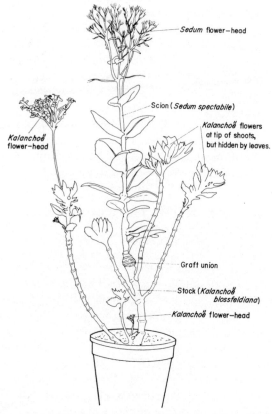

Fig. 95. Transmission of flowering stimulus from short-day plant, *Kalanchoë blossfeldiana*, to long-day plant, *Sedum spectabile*, by grafting. A scion from a vegetative
plant of *Sedum* growing under short days was grafted on to a flowering plant of
Kalanchoë, also under short days. The scion of *Sedum*, as well as the *Kalanchoë* stock,
has flowered under these conditions. (Original by I. D. J. Phillips from plant supplied
by J. Hillman.)

SDP over the past 30 years have nearly always given negative results. There have
been certain claims that extracts of *Xanthium* plants can promote flowering, but
this claim has not always been repeatable by other workers. On the other hand,
flower-promoting, gibberellin-like substances have been extracted from LDP
(p. 208).

This failure to isolate the flower hormone from SDP is baffling, but there are several possible explanations, the three most likely ones being as follows:

1. The hypothetical hormone may be present in extremely low concentrations, or it may be very unstable, so that existing methods of extraction and detection are not sufficiently sensitive.
2. Since the naturally occurring growth hormones are relatively simple compounds, it has generally been assumed that the hypothetical flower hormone will also prove to be a fairly simple substance. There is no reason, however, to exclude the possibility that the flower hormone has a complex, unstable molecule, such as protein or nucleic acid. The fact that the stimulus can be transmitted by grafting, but cannot be extracted, is consistent with this hypothesis, since special techniques would be required to isolate it. The same difficulties of extraction are encountered with certain plant viruses. Attempts to detect differences between the proteins of flowering and non-flowering plants of *Xanthium* or other species have given no clear-cut results. Studies are now being carried out to determine whether there are qualitative differences in specific RNAs between flowering and non-flowering plants.
3. There may be no specific flower hormone—the induction of flowering may depend rather upon *quantitative* differences in the level of certain well-known substances occurring in plant tissues.

We have already seen that dramatic morphogenetic changes can be induced in tissue cultures by varying the relative levels of substances such as auxins and kinins in the culture medium (p. 156). Thus, it is argued, the transition from the vegetative to the flowering condition may be controlled by changes in the levels of, for example, certain growth hormones such as auxins, gibberellins and kinins, or by the balance between these hormones. We shall now consider briefly this possibility.

Effects of Gibberellins and Other Growth Hormones on Flowering

Although attempts to regulate flowering through application of auxins have almost all been unsuccessful (apart from the exceptional case of the pineapple, p. 128) it has been found that a number of LDP can be induced to flower under SD by application of GA_3 (Table 3). LDP which respond to GA_3 are typically species which form a pronounced rosette under SD and which show marked internode elongation ("bolting") under LD. When GA_3 is applied to such species growing under SD there is a very marked stimulation of internode elongation and this process is accompanied by flower initiation.

Gibberellin is also effective in stimulating flowering in the "long-short-day" plants *Bryophyllum crenatum* and *B. daigremontianum*, which normally require to be exposed first to LD and then to SD for flower initiation but which will flower under continuous SD if treated with GA_3. Thus, GA_3 substitutes for the LD requirement in these species.

A number of species which normally require vernalizing before flower-initiation will occur can also be induced to flower by external application of GA_3 (Table 3, Fig. 96). Thus, in these species GA_3 apparently replaces the chilling requirement. However, GA_3 will apparently not stimulate flowering of unvernalized rye and certain other species, although it will stimulate stem elongation in these species. In general, treatment of seeds with GA_3 is not effective in stimulating flowering, even in species which respond to seed vernalization.

TABLE 3. SPECIES SHOWING FLOWERING UNDER NON-INDUCTIVE CONDITIONS IN RESPONSE
TO APPLIED GIBBERELLIC ACID (GA_3)

A. *Long-day Plants*

Anagallis arvensis (Pimpernel)	*Petunia hybrida* (Petunia)
Cichorium endivia (Chicory)	*Raphanus sativus* (Radish)
Hyoscyamus niger (Henbane, annual)	*Rudbeckia bicolor* (Coneflower)
Lactuca sativa (Lettuce)	*Silene armeria*
Papaver somniferum (Poppy)	*Spinacia oleracea* (Spinach)

B. *Plants with Chilling Requirement*

Apium graveolens (Celery)	*Digitalis purpurea* (Foxglove)
Avena sativa (Oat)	*Hyoscyamus niger* (Henbane, biennial)
Beta vulgaris (Sugar beet)	*Matthiola incana* (Stock)
Bellis perennis (Daisy)	*Myosotis alpestris* (Forget-me-not)
Brassica oleracea (Cabbage)	*Solidago virgaurea* (Golden rod)
Daucus carota (Carrot)	

These observations raise the question of whether the endogenous gibberellins are not the "flower hormone" in LDP and species showing vernalization responses. Thus, it might be postulated that in LDP growing under SD the level of endogenous gibberellins is too low for flowering, and that the effect of LD is to raise the level of endogenous gibberellins to the threshold necessary for flowering. Indeed it has been shown for certain species, including spinach and henbane, that there is a marked rise in the levels of endogenous gibberellin when the plants are transferred from SD to LD conditions. Moreover, extracts of gibberellins of the LDP *Rudbeckia*, growing under LD conditions, will induce flowering in plants of the same species growing under SD. Similarly, during vernalization of the biennial species, hollyhock (*Althaea rosea*), there is an increase in the levels of a gibberellin-like substance which will stimulate flowering in *Rudbeckia*, although it will not stimulate flowering in unvernalized hollyhock.

FIG. 96. Effect of vernalization and gibberellic acid on flowering of carrot. *Left:* untreated control plant; *right:* plant chilled for 8 weeks; *centre:* plant unchilled but treated with 10 μg GA₃ per day. (From A. Lang. *Proc. Nat. Acad. Sci. U.S.A.* **43,** 709, 1957.)

Although these latter observations suggest that changes in endogenous gibberellins may be important in flowering, there is other evidence against the hypothesis that flowering in LDP and plants which require vernalization is regulated primarily by gibberellins, including the following:

1. As we have seen, there is evidence that the flowering stimulus is identical in both LDP and SDP, and yet gibberellins are quite ineffective in inducing flowering in most SDP.

2. Not all LDP and species with a chilling requirement can be induced to flower in response to applied GA_3. (However, certain of these latter species can be induced to flower when other types of gibberellin are applied. For example, GA_4 and GA_7 are effective in promoting flowering in *Myosotis alpestris*, which does not respond to GA_3. Similarly, GA_7 is more effective than GA_3 in stimulating flowering in *Silene armeria*. Thus, in some species there appear to be rather precise requirements with respect to the nature of the gibberellin which will stimulate flowering, and these requirements differ between one species and another.)

3. Nearly all rosette plants respond to GA_3 by elongation of the internodes, including those species which do not flower in response to this treatment. Moreover, in henbane, when flowering is induced by LD treatment, the formation of flower primordia *precedes* the elongation of the internodes, whereas in response to GA_3 treatment plants of this species kept under SD begin to show internode elongation *before* the flower primordia appear. These facts suggest that internode elongation and flower initiation are distinct processes and that the primary effect of GA_3 is on internode elongation, with flower-initiation occurring as a secondary effect of stem elongation in certain species.

4. Exogenous GA_3 will not stimulate flowering of "normal" genotypes of red clover (*Trifolium pratense*), but in certain non-flowering genotypes of this species both LD and GA_3 are necessary for flowering. Thus, in these non-flowering genotypes, GA_3 does not substitute for LD, suggesting that some other flower-promoting factor is also normally involved in this species.

Thus, although there is good evidence that changes in the levels of endogenous gibberellins may play an important role in the flowering responses of LDP, it would seem that this is not "the whole story" in LDP and certainly the responses of SDP cannot be accounted for solely in these terms.

In addition to the stimulation of flowering by gibberellins, a number of other growth-regulating substances, both natural and unnatural, have been found to promote flowering in some species under certain conditions. Thus, under certain conditions, kinetin and adenine will promote flowering in *Perilla* and zeatin does so in the aquatic plant *Wolffia microscopica*. Similarly, the naturally occurring growth inhibitor, abscisic acid (p. 234) promotes flowering in *Pharbitis*, *Fragaria* and *Ribes*, while the synthetic growth retardants CCC and B.9 promote flowering in a number of species, including apple and pear trees. A number of other substances, including tri-iodobenzoic acid, maleic hydrazide, vitamin E and even sugars, have been reported to promote flowering in a few species.

Nevertheless, in the majority of SDP and in some LDP, flowering cannot be induced by any combination of the known naturally occurring hormones. Hence we cannot at present account for the flowering behaviour of the majority of species in terms of interaction between known growth hormones.

The conclusion that flowering behaviour cannot be accounted for in terms of the known growth hormones, and yet all attempts to extract a specific flower hormone have been unsuccessful, may indicate that we have adopted an over-simplified approach to the problem, in assuming that flowering is controlled by a single, specific hormone. The isolation and identification of the flowering stimulus remains one of the most challenging problems in developmental plant physiology.

SEX EXPRESSION AND GROWTH HORMONES

There is some evidence that growth hormones may be involved in the determination of sex in some plants. This has come from studies of dioecious species (male and female flowers on *separate* plants) such as hemp (*Cannabis sativa*), and of those monoecious species in which the male and female organs, stamens and ovaries, are borne in separate flowers on the same plant (e.g. some varieties of cucumber (*Cucumis sativus*)). Treatment of genetically male plants of hemp with an auxin spray causes female flowers to be produced. In monoecious cucumber varieties, it is normally the case that male flowers develop during the earlier stages of growth and that female flowers form only later on. However, application of an auxin to the leaves of young cucumber plants results in an acceleration of the transition from production of male flowers to production of female flowers. It has been suggested, therefore, that female flowers or female parts of flowers tend to differentiate under conditions of higher auxin concentration than do male flowers or parts. This conclusion is supported by the finding that genetically-determined forms of cucumber which bear only male flowers contain lower levels of endogenous auxin than the normal hermaphrodite forms. Femaleness in cucumber is also enhanced by treatment with ethylene, or ethrel (a commercial preparation of 2-chloroethane-phosphonic acid, which is converted to ethylene in plant tissues, p. 130). Gibberellin treatment of monoecious cucumber plants, in contrast to auxin treatment, increases the number of male flowers formed. Treatment of gynoecious cucumber (i.e. a dioecious variety which normally produces only female flowers) with gibberellin results in the formation of male as well as female flowers. Moreover, endogenous gibberellin levels are lower in the gynoecious types than in the normal hermaphrodite forms. It is possible, therefore, that sex expression in plants is effected by a balance between endogenous auxins and gibberellins. The effect of auxin on flower sexuality may involve the participation of ethylene (see p. 78).

CHANGES OCCURRING IN THE SHOOT APEX DURING FLOWER INITIATION

We have seen that the transition from the vegetative to the reproductive condition involves drastic changes in the structure of the shoot apical meristems (p. 45). The earliest steps in this transition have been studied in a number of SDP and LDP. Indeed, SDPs such as *Xanthium* or *Chenopodium*, and LDPs such as *Lolium temulentum*, in which flowering may be induced by exposure to a single inductive cycle provide very favourable material for studying the transition, since the timing of the latter can be rigorously controlled to within a few hours. Studies on *Xanthium* have revealed that the earliest changes leading to flower initiation can be first detected about 4 days after exposure to a single SD cycle, but when the plants are exposed to two SD cycles changes can already be detected at the end of this treatment. As we have seen (p. 45), the first changes occur in the region between the central mother cells and the rib meristem, and involve cell division in this region. A stage is soon reached in which a 'mantle' of small, densely staining and actively dividing cells overlies a central core of more vacuolated cells (Fig. 27).

It is clear that when a plant changes from the vegetative to the flowering phase many genes must be brought into action, including those which control flower and fruit development. Thus, we must postulate some mechanism which regulates the "switching on" of the flowering genes. We know very little concerning the nature of such gene-switching mechanisms in higher plants, but this subject is discussed further in Chapter 13. Since genes, i.e. DNA, control development by the synthesis of specific enzymes, through a process involving various types of RNA (p. 274),† it is evident that nucleic acid metabolism is likely to be very intimately involved in the events occurring at the shoot apex during flower initiation. The flowering stimulus, formed in the leaves of photoperiodic plants, must, therefore, act at the shoot apex by affecting the gene-switching processes and thus involve nucleic acid metabolism.

Nucleic acid changes occurring at the shoot apex during flower initiation have been studied in two ways: (1) by following the incorporation of radioactive precursors into RNA, and (2) by the use of inhibitors of RNA synthesis.

In the LDP, *Lolium temulentum*, there is marked incorporation of radioactive precursors into RNA (thus indicating active RNA synthesis) at the shoot apices on the morning of the day following a single LD cycle, which is precisely the time at which the flowering stimulus is estimated to arrive at the apex. These changes in RNA and protein synthesis are most prominent in the cells on the flanks of the meristem which are destined to give rise to the spikelets (Fig. 97). Similarly, in the LDP, *Sinapis alba* (mustard), which will also flower in

† Readers who are not familiar with this process are advised to read pp. 273–275 at this point.

response to a single LD cycle, there is a marked increase in RNA synthesis in the central and peripheral zones of the apical meristem at about 17 hours following the beginning of the LD cycle. Later, active DNA synthesis occurs, and this is followed by mitosis.

The need for RNA and DNA synthesis has also been studied by using anti-metabolites, such as 2-thiouracil (2TU), which inhibits RNA synthesis, and

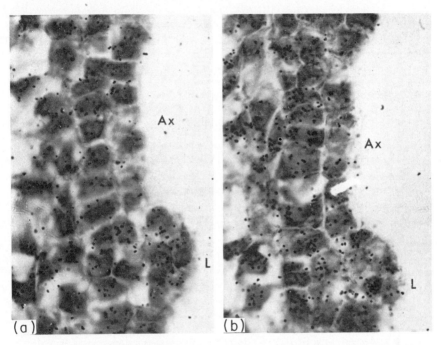

Fig. 97. Autoradiographs of vegetative and induced shoot apices of *Lolium temulentum*, labelled with [³H] orotic acid. A. Axillary bud site (Ax) and leaf primordium (L) in a vegetative (short-day) apex. B. Axillary bud site of plant which has been exposed to long day, in which there has been active incorporation of orotic acid (as shown by the density of silver grains), indicating active nucleic acid metabolism. (From R. B. Knox and L. T. Evans, *Austr. J. Biol. Sci.* **21**, 1083, 1968.)

5-fluorodeoxyuridine (5FDU) which inhibits DNA synthesis. Thus, application of 2TU to the shoot apex of *Sinapis alba* is most inhibitory to flowering when applied during the 12th to 20th hours after the beginning of a single LD cycle, suggesting that RNA synthesis during this period is an essential requirement for flower initiation at the shoot apex. Comparable results have been obtained with the SDP, *Xanthium* and *Pharbitis*. By studying the effects of 5FDU, it also appears that DNA synthesis is essential for flower initiation in these two latter species. However, in *Pharbitis*, 5FDU is effective in inhibiting flower initiation even if applied 24 hours after the end of an inductive dark period, (i.e. after the arrival of the flower hormone) suggesting that it may inhibit the later formation of flower primordia, rather than the processes initiated by the flowering stimulus on its arrival at the shoot apex.

The results of these studies are fully consistent with the view that the initial events associated with the arrival of the flowering stimulus at the shoot apex involve marked changes in the nucleic acid metabolism, which might be expected if there is a "reprogramming" of gene activity during the change from vegetative to reproductive development.

NUTRITION AND FLOWERING

Farmers and gardeners have long known that fertilizers and manures have a marked effect on the balance between vegetative and reproductive growth. Indeed, for many practical purposes there seems to be an antagonism between vegetative growth and reproduction, and manurial treatments which favour strong vegetative growth may be unfavourable for flowering and fruiting. Experimental studies give some support for this view. Thus, it is found that low levels of nitrogen tend to result in earlier flowering in certain long-day plants. It has been shown that high nitrogen and carbohydrate nutrition delay flower initiation in pea (*Pisum sativum*). However, the number of cases in which a clear effect of mineral nutrition on the onset of flowering has been demonstrated is rather small. Mineral nutrition appears to have an important effect on flower initiation in fruit trees, high levels of nitrogen tending to promote vegetative growth and reduce flowering.

FLOWERING IN "NEUTRAL" SPECIES

In a large number of species the onset of flowering is a day length or chilling response, and the discovery of photoperiodism and vernalization represents a very real advance in our understanding of the physiology of flowering. However, it must be remembered that in many other species, probably equally numerous, flowering is not greatly affected by daylength or winter-chilling (Table 4). This is the group referred to earlier (p. 165) in which flowering is relatively insensitive to external conditions, and although in these species the length of the vegetative phase may be *modified* by environmental conditions, the effects of the latter are not completely overriding. However, such species, which will be referred to as "neutral" species, are not sharply distinguished from those showing a "quantitative" response to daylength, which have been referred to as "facultative" LDP or SDP (p. 167).

In these neutral species, flowering appears to be determined primarily by some *internal* mechanism, since even when grown under constant environmental conditions a species such as sunflower will grow vegetatively for a certain period and then become reproductive; thus the transition cannot be caused by any change in external conditions and must be regulated by some "internal" mechanism. Moreover, it would appear that the transition from the vegetative to the flowering condition in such species is but one manifestation of a more general phenomenon, since progressive changes during development are very common, as shown by the development of morphological differences in successive organs, such as leaves. It is commonly found that there are changes in the size and shape of successive leaves, the seedling or primary leaves being smaller than the later ones, and in species with deeply indented or compound mature leaves the primary leaves are usually much simpler in form (Fig. 98) with a progressive series of later formed leaves showing increasing segmentation. Plants showing such changes are said to exhibit *heteroblastic development*. These changes may be affected by environmental factors, such as light intensity and mineral nutrition, but they are not *dependent* on environmental changes and will occur even under constant conditions.

TABLE 4. SOME EXAMPLES OF SPECIES SHOWING "NEUTRAL" FLOWERING RESPONSES

Cucumis sativus (Cucumber)	*Phaseolus vulgaris* (Dwarf bean)
Fagopyrum tataricum (Buckwheat)	*Poa annua* (annual meadow grass)
Fuchsia hybrida (Fuchsia)	*Rosa* spp. (Rose)
Helianthus annuus (some varieties)	*Senecio vulgaris* (Groundsel)
Lathyrus odoratus (Sweet pea)	*Solanum tuberosum* (Potato)
Lycopersicum esculentum (Tomato)	*Vicia faba* (Broad bean)
Nicotiana tabacum (Tobacco, certain varieties)	

The changes in leaf shape appear to reflect progressive changes in the size and shape of the shoot apex. For example, in the sunflower plant the diameter of the subapical region increases progressively as the plant grows and ultimately these changes are terminated by the formation of an inflorescence. It seems likely that the changes in leaf shape result from these changes in the shoot apex, leaf primordia produced on a larger apex apparently being capable of continuing their development for a longer period and hence producing a more mature leaf form.

This heteroblastic development appears to be an indication of the progress of the plant towards maturity and the attainment of the reproductive condition. For example, early flowering varieties of the cotton plant (*Gossypium*) show a steep gradient in leaf shape changes, whereas in later-flowering types the rate of change in leaf shape is less steep.

Evidence of progressive changes towards the flowering condition has come from experiments in which internode segments of tobacco were taken and grown in sterile culture. These segments produced callus and regenerated buds.

Internode segments from young plants or from the lower region of the stem of older plants produced vegetative buds, whereas stem segments from the upper part of a flowering plant produced flower buds, with few leaves or bracts. Segments taken from an intermediate region of the stem first produced leaves and bracts and then a flower. Thus, there appeared to be a gradient in the propensity to produce flowers, from the lower to the upper part of the plant.

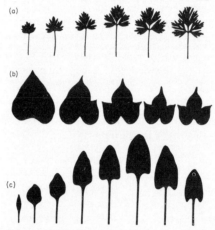

Fig. 98. Heteroblastic development, as shown by successive leaves of (a) *Delphinium ajacis*, (b) morning glory (*Ipomoea hederacea*), (c) sugar beet (*Beta vulgaris*). (From E. Ashby, *New Phytologist*, **47**, 153, 1938.)

The nature and causes of the changes occurring in the shoot apex during development are completely unknown but there would appear to be at least three general possibilities:

1. There is some inherent pattern of behaviour of the apex, which will pursue its appointed course, independently of the differentiated portions of the plant.
2. The behaviour of the apex is affected and determined by influences arising in the mature parts of the plant, e.g. it might depend upon the attainment of a certain minimum leaf area.
3. The behaviour of the apex may be dependent upon the gradual accumulation of certain metabolites which must attain a certain threshold concentration before flowering will occur.

At present, there is no incontrovertible evidence to enable us to decide which of these hypotheses is correct. Attempts have been made to obtain evidence on this problem in experiments with sunflower, in which seedlings tips were grafted on to stocks of various ages and tips from older plants were grafted on to seedlings. In the first type of graft it was found that seedling tips formed fewer nodes than they would have done on their parent plants. On the other hand, grafts of tips from older plants on to seedlings formed a greater number of nodes than they would have on the parent plants. These results would seem to indicate that the behaviour of the apex is determined by influences from the mature parts of the plant.

It will be remembered that among plants which are sensitive to daylength or chilling, there are some species which will not respond until they have reached a certain minimum size ("ripeness-to-flower"), and before they have reached this stage seedling plants may be described as in a "juvenile phase". This phenomenon would seem to correspond to the necessity for neutral species to reach a certain size before flowering. Indeed, photoperiodic and vernalizable species will behave like neutral species if they are maintained from germination under constant conditions favourable to flowering, in that they will undergo a certain period of vegetative growth and then commence to initiate flowers. Thus, it would appear that the physiology of flowering is not qualitatively different in the "sensitive" and "neutral" groups of plant, but rather that they differ in *degree* of sensitiveness to external conditions.

If we are correct in assuming that the difference between the flowering responses of "sensitive" and "neutral" species is primarily one of degree, then it might be assumed that a flower hormone is involved in neutral as well as in sensitive species, but there is little evidence as yet for the occurrence of flower hormones in neutral species.

THE DIVERSITY OF FLOWER-CONTROLLING FACTORS

As we have now seen, flowering is affected and controlled, in various species, by a range of factors, some of which are external to the plant, such as daylength, temperature and nutrition, and others which arise within the plant itself, as seen in day-neutral species. There is a bewildering mass of facts on the effects of these various factors and it is difficult to see how these parts of the "jigsaw puzzle" fit together to give an overall, unified picture. It might be asked whether the fact that quite different environmental factors may control flowering implies a different control mechanism for each factor. In attempting to answer this question it is useful to approach the problem from a different viewpoint and to ask not, what makes a plant flower under a given set of conditions, but rather, why they *fail* to flower under a different set of conditions. Thus, flowering may be prevented by a number of different factors in various species, although the flower-promoting processes might be the same in these species.

For example, in apple trees, flower initiation is rather sensitive to mineral nutrition, but not to photoperiod. We do not need to postulate that flower hormone synthesis depends upon low nitrogen levels—it may well be the case that the "flower hormone" is always present in apple shoots, but that under conditions of high nitrogen nutrition there are high levels of, for example, endogenous gibberellins, which are known to inhibit flowering in this plant. On the other hand, in *Xanthium* it would appear that synthesis of the flower hormone is blocked under long days. In plants requiring vernalization, yet another step in the flowering processes may be blocked in unchilled plants. From this viewpoint, under favourable conditions the flowering processes may be blocked in different ways and at different steps in plants of various response types. Similarly, as we have seen, flowering may be blocked by different factors in the same plant at different stages of its life cycle. In the seedling stages flowering may be prevented because the plant is still in a juvenile phase and is not yet capable of responding to favourable environmental conditions. But even when it has attained ripeness to flower, it may be prevented from flowering by unfavourable environmental conditions.

It is not difficult to envisage that in some cases flowering does not occur because the stimulus is not being synthesized in the leaves, while in other cases the limiting step may be the inability of the shoot apices to respond to the stimulus, possibly because of inappropriate levels of gibberellins.

FLOWERING IN WOODY PLANTS

So far, our consideration of the physiology of flowering has been restricted to herbaceous plants. Flowering in woody plants presents a number of characteristic features which will be briefly described.

The first important feature to note is that there is considerable variation between species in the length of time elapsing between flower initiation and complete development of the flower. In some species, such as sweet chestnut and many other late summer flowering trees and shrubs, such as *Buddleia*, *Fuchsia*, *Hypericum*, *Hibiscus syriacus*, *Caryopteris*, etc., the flowers are formed on the current year's shoots and flower initiation is followed immediately by the further full development of the flower, as in herbaceous species, i.e. there is no gap between the early and later stages of flower development. In many other woody plants, however, especially temperate species, flower initiation occurs during the summer within resting buds formed earlier in the same year, but the development of the flower parts becomes arrested at an early stage and the further development and emergence of the flower does not occur until the following spring. This is the situation in a large number of common European woody plants, e.g. oak (*Quercus*), ash (*Fraxinus*), sycamore (*Acer*), elm (*Ulmus*), pine (*Pinus*), apple, plum, peach, black currants, gooseberries, etc. In such species the buds containing flower primordia become dormant in the late summer or autumn and require a period of exposure to chilling temperatures to break the dormancy (p. 230). Once the dormancy has been broken by winter chilling they become capable of emerging as the temperatures rise in the spring.

In some winter or early spring flowering trees and shrubs the buds containing flowers are able to grow at lower temperatures than the vegetative buds, so that the flowers may actually emerge before the leaves, e.g. hazel (*Corylus*), willow (*Salix*), jasmine (*Jasminium nudiflorum*), almond (*Prunus persica*), elm, ash, oak, etc.

We have very little information regarding the effect of environmental factors on flower initiation in woody plants. This is mainly due to the technical difficulties in experimenting with mature trees. It is not possible to work with seedling trees since these are still in the juvenile phase and are not capable of flowering (p. 220). However, it is possible to study the effects of environmental conditions on flowering of trees by taking scions from mature trees and grafting them on to seedling stocks. In this way it is possible to obtain small trees which are potentially capable of flowering. In a few woody plants, flower initiation appears to be controlled by daylength conditions. Thus, long days are necessary for flower initiation in birch (*Betula*), *Erica* and *Calluna*. On the other hand, short days promote flower initiation in coffee (*Coffea arabica*), black currant (*Ribes nigrum*), *Poinsettia* and *Hibiscus*. Flower initiation in other woody plants, e.g. pine, larch, beech, apples, cherries, plums, appears to be unaffected by photoperiodic conditions, which, nevertheless, have a profound effect on vegetative growth of some of these species (p. 227).

In general, vernalization does not appear to play an important role in flower initiation in woody plants, with a possible exception of the olive tree (*Olea europaea*), which apparently initiates flowers in response to cool temperatures in the winter.

On the other hand, there is no doubt that temperature, rainfall and soil nutrients may have a profound effect on the flowering of many woody plants. It is well known that a hot, sunny summer is frequently followed by abundant flowering of many tree species in the following spring, and the flowering of beech has been shown to be particularly affected in this way. Evidently high temperatures and possibly high light intensity (and hence the formation of abundant carbohydrate reserves) favour flower initiation in many tree species. The effect of mineral nutrition on flower initiation in fruit trees was mentioned earlier (p. 214).

A number of instances have been reported in which flower initiation in woody plants has been influenced by gibberellins and by synthetic growth retardants. Thus, "flowering" has been stimulated by gibberellic acid in a number of conifers, including *Cupressus*, *Chamaecyparis*, *Juniperus* and *Thuja*. By contrast, gibberellic acid inhibits flowering in apple, pear, black currant, grape (*Vitis cimifera*), *Syringa*, *Fuchsia*, and a number of other woody plants. On the other hand, growth retardants such as "CCC" (Chlorocholine chloride) and "Phosfon D", which apparently inhibit gibberellin biosynthesis in plant tissues promote flowering in apples, pears and azaleas (*Rhododendron* spp.).

PHASE CHANGE IN WOODY PLANTS

We have seen that in "day-neutral" annual species the onset of flowering is apparently regulated by some endogenous mechanism, the nature of which is unknown, but that the effect of this regulatory mechanism is that the plant does not flower until it has attained a certain size. This "size effect" is seen also in species showing daylength or chilling responses, so that there is a juvenile phase during which flowering cannot be induced. An apparently analogous phenomenon is seen in woody plants, the seedlings of which normally show a juvenile phase during which they make active growth but remain vegetative The transition to the flowering condition occurs only after a delay which varies greatly from species to species, ranging from 1 year in certain shrubs to 30–40 years in forest trees such as beech (Table 5). Once a given tree commences

TABLE 5. DURATION OF JUVENILE PERIOD IN FOREST TREES

	years
Pinus sylvestris (Scots pine)	5–10
Larix decidua (European larch)	10–15
Pseudotsuga taxifolia (Douglas fir)	15–20
Picea abies (Norway spruce)	20–25
Abies alba (Silver fir)	25–30
Betula pubescens (Birch)	5–10
Fraxinus excelsior (Ash)	15–20
Acer pseudoplatanus (Sycamore, Maple)	15–20
Quercus robur (English oak)	25–30
Fagus sylvatica (Beech)	30–40

flowering, it normally continues to do so every year, although, as we have already seen (p. 219), in species such as beech, flowering is sensitive to weather conditions and may occur irregularly. Thus, on the basis of the flowering behaviour we may distinguish between a *juvenile* and an *adult* (or *mature*) phase in the life history of the tree.

Differences between juvenile and adult stages are seen not only in flowering behaviour but also in various vegetative characters. Thus, in certain species, such as ivy (*Hedera helix*), mulberry (*Morus*), *Acacia*, *Eucalyptus* and juniper, the leaf shape in the juvenile phase is very different from that of the adult stage. In oak and beech there is a marked tendency for the dead leaves to be retained on the shoots of juvenile trees during the winter, whereas in the adult stages they are shed normally. In some species the juvenile and adult stages are distinguished by marked differences in phyllotaxis, as in ivy (see below). Another morphological character which changes during ontogeny is the development of thorns—for example, in lemon trees the juvenile stages are commonly more thorny than the adult.

Among the physiological differences observed between juvenile and adult stages is the rooting ability of cuttings; it is very commonly found that whereas cuttings from young trees root readily, after the parent tree attains a certain age this rooting ability is greatly diminished or entirely lost.

FIG. 99. Cuttings taken from adult part of parent vine of ivy (*Hedera helix*). These cuttings will continue to retain for several years the "adult" characters, such as leaf shape, phyllotaxis, and the capacity to flower, although high temperatures favour reversion to the juvenile condition. (Print supplied by Dr. L. W. Robinson.)

An interesting feature of these phase differences is that the lower parts of the tree retain the juvenile state after the upper parts have developed adult characters. This can be seen very well in ivy, where the lower regions of an erect-growing vine show the palmate type of leaf and are purely vegetative, whereas the upper parts show ovate leaves and normally produce abundant flowering shoots.

Not only is there apparent stability of the juvenile and adult phases in the same plant body, but cuttings taken from different regions and rooted retain their juvenile or adult characters for a long period. This is well known in ivy, for example, where cuttings taken from the juvenile part of the vine have palmate leaves with opposite phyllotaxis, the shoots are trailing and are coloured with anthocyanin and produce abundant adventitious roots, but do not flower; cuttings of adult shoots, on the other hand, produce plants with ovate leaves, spiral phyllotaxis, and the shoots are erect, green and produce few, if any adventitious roots, but flower readily (Fig. 99). Cuttings from adult ivy shoots may grow for many years and produce shrubs known to gardeners as "tree ivy".

A similar retention of juvenile and adult characters is seen in grafting experiments. Thus, scions from flowering regions of mature trees continue to flower when grafted on to quite small seedling stocks. This is a common observation in forest tree breeding and is known for species such as birch, larch and pine. Likewise, scions from mature fruit trees grafted on to suitable root stocks flower readily, whereas scions from young seedlings treated in the same way show delayed flowering.

The phenomenon of phase change, involving stable, non-genetic changes which can be transmitted through many cell-generations, shows certain interesting parallels with the changes occurring during vernalization. This subject will be discussed further in Chapter 13.

FURTHER READING

General

EVANS, L. T. (Ed.). *Induction of Flowering*, MacMillan, New York and London, 1969.

HENDRICKS, S. B. How light interacts with living matter. *Sci. Amer.* **219**, 175, 1968.

HILLMAN, W. S. *The Physiology of Flowering*, Holt, Rinehart & Winston, London and New York, 1964.

LAETSCH W. M. and R. E. CLELAND (Ed.). *Papers on Plant Growth and Development*, Little, Brown & Co., Boston, 1967.

SALISBURY, F. *The Flowering Process*, Pergamon Press, Oxford, 1963.

WILKINS, M. B. (Ed.). *The Physiology of Plant Growth and Development*, McGraw-Hill, London, 1969.

More Advanced Reading

CHOUARD, P. Vernalization and its relations to dormancy. *Ann. Rev. Plant Physiol.* **11**, 191 1960.

DOORENBOS, J. Juvenile and adult phases in woody plants. *Encycl. Plant Physiol.* **15**(1), 1222, 1965.

LANG, A. Physiology of flower initiation. *Encycl. Plant Physiol.* **15**(1), 1380, 1965.

PURVIS, O. N. The physiological analysis of vernalization. *Encycl. Plant Physiol.* **16**, 76. 1961.

CHAPTER 11

Dormancy

OUTSIDE the equatorial regions there are seasonal variations in climatic conditions, which are most notable in the temperate zones. These variations are especially marked with respect to light intensity, daylength, temperature and frequently also to rainfall. As a result there is a regular alternation of seasons favourable and unfavourable for growth and this alternation has had a marked effect on the pattern of the life cycles evolved by the higher plants. The necessity to withstand low temperatures during the winter, and, in some regions, hot dry conditions during the summer, poses special problems for the plant, and we now have to consider some of the ways in which these problems have been met.

THE BIOLOGICAL SIGNIFICANCE OF DORMANCY

Plant cells normally contain a large amount of water which is liable to freeze at low temperatures, with grave risk of damage to the protoplasm. Tropical plants are very easily killed by frost, but it is evident that plants of temperate and arctic regions must have become adapted to survive the period of winter frost—they have developed *frost-resistance*. Although frost-resistance has been studied for many years, our understanding of its biochemical basis is still far from complete, and a discussion of this subject here would take us too far afield.

In many frost-resistant species the general morphological appearance of the plant during winter is not essentially different from that in the summer—the growth rate of the plant is reduced or arrested during the winter, but the growing points of the shoots are still in a potentially active condition, and may make some growth during mild periods, as in many biennial plants. In such species the whole plant, including the apical meristems, is relatively frost-resistant. Other species, of course, show distinct differences between their summer and winter states. Thus, in woody plants the shoot apices cease active

223

growth and become enclosed in bud scales, to form winter resting buds. They are then said to have become *dormant*. Many woody plants are much more frost resistant in the dormant than in the actively growing condition. Thus, seedlings of forest trees, such as larch (*Larix*) and *Robinia*, which continue growing late into the autumn, are very liable to be damaged by early frosts, but if they have ceased growth and their growing points have entered the dormant condition, they then remain frost-resistant throughout the winter.

The reason why dormant buds should be more frost-resistant than actively growing tissues is not fully understood. However, it is fairly clear that the frost-resistance of dormant tissue is due to certain protoplasmic characters, and that it is not primarily due to the presence of the bud scales, the protective function of which is probably concerned with the reduction of water loss—one of the secondary effects of winter cold is to increase the difficulties of the plant in maintaining an adequate water balance. Under frosty conditions, especially when accompanied by wind, the plant continues to lose water, but is unable to take up replacement supplies if the soil is frozen. There is thus a considerable danger of damage from drought under winter conditions, but water loss is reduced by the enclosure of the growing parts within a covering of bud scales, and also, in deciduous trees, by the falling of leaves in autumn which reduces the total surface area over which evaporation can occur.

The danger of winter drought and low temperature seems to have influenced not only the evolution of woody plants, but has apparently had a profound effect on the form of many other types of plant. Many plants over-winter entirely below ground, as bulbs, corms and rhizomes; although such organs will be partly insulated against frost, they will also be protected against drying when the soil does become frozen to some depth. Some dormant organs, such as bulbs, are probably adaptations to hot, dry summer conditions, such as are found in the Mediterranean region.

Whereas perennial plants have developed special organs which resist the unfavourable conditions of winter, annual plants have pursued yet another course—they frequently over-winter in the form of seeds. The seeds of many annual plants, particularly of the common weeds of arable land, germinate almost immediately they are shed, if conditions of temperature and moisture are favourable. But the seed of many other plants does not germinate imme-diately (or only a proportion of the seeds do so) and remains in the soil until conditions become favourable for germination in the following spring. Now seeds are generally very much more frost resistant than the growing plant of the same species. Dry seeds may resist freezing down to as low as $-234°C$. Some seeds, such as those of Leguminosae (clover (*Trifolium*), broom (*Cytisus*), *Laburnum*, etc.) do not, in fact, take up water immediately they are shed, due to the fact that they have a coat which is impermeable to water, and such seeds will be capable of withstanding very severe frost.

The majority of seeds imbibe water as soon as they fall on to moist soil, but, as already stated, they do not necessarily germinate immediately. Such imbibed seeds are less frost-resistant than in the dry state, but nevertheless many of them still retain a considerable degree of frost resistance, and apparently certain annual species which are frost-tender in the actively growing state are able to survive the winter in the form of seed.

FORMS OF DORMANCY

Dormancy may be defined as a state in which growth is temporarily suspended. In some species the cessation of growth is directly due to unfavourable temperature and light conditions; thus many pasture grasses remain in continuous growth throughout a mild winter and cease growth only when temperatures fall to about 0–5°C. Similarly, certain annual weeds, such as groundsel (*Senecio vulgaris*), chickweed (*Cerastium* spp.) and Shepherd's purse (*Capsella bursa-pastoris*), stop growing only during the coldest part of the winter. In such cases the dormancy of plants is evidently caused by the unfavourable external conditions, and in this case we speak of *imposed* or *enforced dormancy*.

However, in many cases the unfavourable conditions are not directly the cause of dormancy. Thus, many trees form winter resting buds during the summer and autumn, when temperatures and light conditions are still favourable, and long in advance of the onset of winter. In such woody plants the cause of dormancy appears to lie within the tissues of the buds themselves, and we then speak of *innate* or *spontaneous dormancy*. This form of dormancy also occurs in many seeds. Thus, if freshly harvested barley grains are planted under warm, moist conditions, a high percentage of them will fail to germinate. If, however, the barley is stored dry for a few months, the seed will then be found to germinate readily when planted under the same conditions as previously. Thus, the failure of freshly harvested barley grains to germinate is not due to external conditions being unfavourable for growth, but must be due to some cause within the seed itself.

Innate or spontaneous dormancy is found not only in buds and seeds, but in other types of resting organs such as rhizomes, corms and tubers.

BUD DORMANCY IN WOODY PLANTS

The majority of temperate woody plants, including both coniferous and dicotyledonous species, show a well-marked dormancy or resting phase during the annual growth cycle and this is usually accompanied by the development of resting buds. The typical resting bud involves the "telescoping" of the bud scales and leaf primordia in the apical region, due to the arrest of normal internode extension. In some genera (e.g. *Betula*, *Fagus*, *Quercus*) having stipules, this telescoping of the shoot apical region leads to the formation of a resting bud, since the overlapping stipules in this region form the bud scales. In other species the protective scales represent leaves, which may be only slightly modified, as in *Viburnum* spp., or more highly modified, so that they frequently represent only the leaf base, as in *Acer*, *Fraxinus*, *Malus* and *Ribes* (Fig. 100).

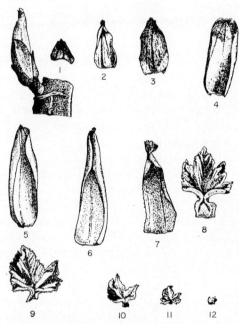

Fig. 100. Bud of *Ribes* and bud scales and leaves from a dissected bud. (From J. H. Priestley and L. I. Scott, *An Introduction to Botany*, Longmans, Green & Co., 3rd ed., London, 1955.)

During the development of such buds, certain leaf primordia show greater marginal growth than occurs during normal leaf development, whereas lamina development is suppressed, and these primordia give rise to the bud scales. The younger leaf primordia, formed within the bud scales, are arrested at an early stage in their development, and give rise to normal leaves when the buds resume growth in the following spring. In some species, such as pines, growth of a bud may continue for several months from June to September. In some trees a terminal bud is not formed, since growth of the shoot is terminated by the death and abscission of the apical region, and growth is later continued from the uppermost axillary bud. Such species (which include

Tilia, Ulmus, Castanea, Robinia and *Ailanthus*) are said to show a *sympodial* (as opposed to a monopodial) growth habit.

When terminal buds are first formed they can frequently be induced to resume growth by various treatments including defoliation, either by hand or by insect attack. It would appear, therefore, that at this stage the terminal buds are not themselves innately dormant, but their growth is apparently inhibited by the mature leaves on the shoot. Similarly, lateral buds may be inhibited by the leaves, or by the main apical region in actively growing shoots, and are thus held in check by correlative inhibition (p. 122) rather than by innate dormancy. This phase of bud development is referred to as *summer dormancy* or *predormancy*. Later, in many species, the buds are found to have entered a state referred to as *true dormancy, winter dormancy* or *rest*. When they have entered this condition the buds will no longer resume growth if the shoots are defoliated, so that they are now innately dormant and not simply held in check by environmental conditions or inhibitory influences within the plant itself, as is the case during predormancy.

After a certain period of true dormancy the buds become capable, in the later part of the winter or early spring, of resuming growth when external conditions, particularly temperature, are favourable for growth. Thus, at this stage the buds are no longer innately dormant, but nevertheless for some time they may fail to grow because of low outdoor temperatures. This phase is referred to as *post-dormancy*. We shall now consider some of the environmental factors controlling the development and breaking of dormancy in temperate woody plants, of which daylength and temperature are the most important.

The Development of Bud Dormancy

One of the most important factors affecting and controlling the induction of dormancy in woody plants is daylength. In the majority of species so far studied, long days promote vegetative growth and short days bring about the cessation of extension growth and the formation of resting buds in seedlings of woody plants (Fig. 101). However, a number of common cultivated fruit trees (*Pyrus, Malus, Prunus*) and certain other species, including the family Oleaceae, appear to be relatively insensitive to daylength changes.

The seedlings of some species, e.g. black locust (*Robina pseudacacia*), birch (*Betula pubescens*) and larch (*Larix decidua*), can be maintained in continuous growth for at least 18 months under LD conditions in a warm greenhouse, whereas under SD they cease growth within 10–14 days. On the other hand, other species such as sycamore (*Acer pseudoplatanus*), horse chestnut (*Aesculus hippocastanum*) and sweet gum (*Liquidambar styraciflua*), show delayed dormancy under LD, but they cannot be maintained in growth indefinitely under these conditions. For those species which can be maintained in continuous growth under LD, there appears to be a certain critical daylength below which dormancy is induced and above which dormancy does not occur.

Fig. 101. Photoperiodic control of bud dormancy in seedlings of birch (*Betula pubescens*). Plants transferred to short days (left) have ceased growth and formed resting buds, whereas seedlings maintained under long days will continue to grow actively for many months.

As in the flowering responses of herbaceous plants, the photoperiodic responses of woody seedlings appear to depend upon the length of the dark period, rather than of the photoperiod, and if a long dark period is interrupted by a short "light-break", the effect of the dark period is nullified and dormancy is delayed. The most effective region of the spectrum for this light-break effect lies in the red, suggesting that phytochrome is involved, and clear red/far-red reversibility has been demonstrated for seedlings of larch.

The response of woody seedlings depends on the daylength conditions to which the leaves are exposed. In sycamore, as in herbaceous species, it is the young, fully expanded leaves which are the most sensitive to daylength, but in birch seedlings even quite young leaves in the apical regions show sensitivity to photoperiod.

How important are these photoperiodic responses in determining the formation of resting buds and the onset of dormancy in nature? It has been shown that the seasonal decline in daylength is important in determining the onset of dormancy in seedlings of species which normally continue active growth into the autumn, e.g. *Larix decidua*, *Populus* spp. and *Robinia pseudacacia*. But it is very common to find that older trees show a very much shorter period of extension growth than do seedlings of the same species, and they frequently cease growth in June or July, when the natural photoperiods are still long. In such cases, it seems doubtful whether declining daylength is important in determining the formation of resting buds, and it seems more likely that some change, in either nutrient levels or in hormonal balance, arising within the tree itself, determines the period of growth and the onset of dormancy. However, as we have seen, resting buds are at first in a state of predormancy and only later enter a state of true dormancy; it is possible that the declining daylength in the autumn plays a role in the transition of the buds from predormancy to true dormancy.

It has been found that leaf fall in some woody plants is promoted by short-days, and delayed leaf fall is sometimes observed in trees growing near street lights. However, in nature, low temperatures and possibly low light intensity, are probably at least as important as daylength in determining the onset of leaf senescence and abscission.

In addition to the observable morphological changes associated with the induction of dormancy by short-days there are also biochemical changes which are reflected in increased cold resistance. Seedlings of black currant (*Ribes nigrum*) and *Robinia pseudacacia* which have been exposed to SD are markedly more frost-resistant than seedlings which have been grown throughout under LD. Cold acclimatization in *Cornus stolonifer* depends upon exposure to both SD and decreasing temperature.

It is now well established that wide-ranging woody species, such as *Pinus sylvestris* and *Picea abies*, show marked ecotypic differences in photoperiodic responses in relation to both the latitude and the altitude at which they occur naturally. Northern races are frequently found to require longer photoperiods for active extension growth than do more southern races, adapted to shorter natural photoperiods. This fact suggests that woody plants are rather closely adapted to natural daylength conditions, and that the latter probably play an important controlling role in the seasonal cycle of growth and dormancy.

Emergence of Buds from Dormancy

Usually the terminal resting buds formed during the summer or fall remain dormant until the following spring, when they expand and form new shoots. The dormancy of the buds diminishes during the course of the winter, as can be demonstrated very simply by collecting twigs of trees such as lime (*Tilia*), sycamore (*Acer pseudoplatanus*), poplar (*Populus*) and willow (*Salix*) at different times during the winter and placing them in water in a warm room or greenhouse. It is found that twigs collected in October, November and early December usually remain dormant when they are brought into warm conditions. A fairly high proportion of buds collected in January will be found to expand in 2–3 weeks, and with later dates of collection, e.g. February or March, the buds burst increasingly rapidly after they are brought into the warm.

Many woody plants require to be exposed to a period of winter chilling to overcome the dormancy of their buds, as can be shown by growing small trees of species such as poplar or sycamore in pots and, when they have become dormant in the autumn, keeping some of them out-of-doors throughout the winter and some of them in a warm room or greenhouse. In the spring, the buds of young trees which have been kept outdoors will expand in the normal way, but those which have been kept in the warm will still be dormant and may remain so until well into the summer; indeed, a certain proportion may ultimately die without ever resuming growth. Temperatures in the range 0°–5°C are the most effective in overcoming bud dormancy, and chilling periods of 260 to 1,000 hours are required. In regions with cold winters, the chilling requirements are normally fully met by the spring, but in warm climates, such as those of California and South Africa, where the winters are very mild, difficulties may be met in cultivating certain fruit trees, such as peaches (*Prunus persica*), since the chilling requirements of the buds may not be met and delayed and irregular bud-break may occur in the spring.

It should be noted that although chilling is necessary to remove the dormancy of the buds of many trees, *warm* temperatures are necessary for the growth of the buds after the chilling period. Frequently the chilling requirements are met by January, but in many regions the buds may fail to resume growth then because the temperatures are still too low, and they remain in the phase of post-dormancy. Thus, the time of bud burst in the spring is normally determined by the return of warmer conditions.

In the majority of North Temperate woody species so far studied, once bud dormancy has been fully induced by SD, they cannot be induced to resume growth by transfer to LD, and the dormancy can normally be overcome only by chilling. However, in a few species the unchilled buds can be induced to resume growth under LD or continuous illumination. Thus, if leafless seedlings of beech (*Fagus sylvatica*), birch (*Betula* spp.) or larch (*Larix decidua*) are placed under continuous light in a warm greenhouse in the autumn, the buds will soon expand.

At first sight it would seem that the response of dormant buds to photoperiod contradicts the rule that it is the leaves which are the organs of photoperiodic "perception", and that the apical meristematic region is insensitive to daylength. However, it should be remembered that resting buds contain well-developed leaf primordia and that therefore the differences between species, such as beech, and other species relates primarily to a difference in the age at which the leaves become sensitive to photoperiod.

It is not clear whether the photoperiodic control of bud-break is important in nature, but there is evidence that bud-break in *Fagus sylvatica* may be dependant upon lengthening daylengths in the spring, although in many regions temperature is also likely to be a limiting factor for this species. In *Rhododendron*, also, bud-break appears to be determined by daylength.

DORMANCY IN VARIOUS ORGANS

Various other types of organ, such as rhizomes, corms, bulbs, tubers and the winter resting buds of aquatic plants, show dormancy. In the aquatic plants *Stratiotes*, *Hydrocharis* and *Utricularia* the dormancy of the winter resting buds is induced by short days, in association with high temperature. Short days also promote the formation of resting buds in the insectivorous plant, *Pinguicula grandiflora* and the dormancy of the buds is overcome by chilling.

By contrast, dormancy in bulbs of onion (*Allium cepa*) is promoted by LD so that the bulbs develop and "ripen" in the summer. The period of dormancy of onion bulbs is shortest when the bulbs are stored at cool temperatures.

The rhizomes of lily-of-the-valley (*Convallaria majalis*) normally become dormant during the summer and they require a one-week period of chilling at 0·5–2°C, or 3 weeks at 5°C to remove this dormancy. Similarly, when *Gladiolus* corms are exposed to warm soil temperatures they do not grow, but periods as short as 24 hours at 0–5°C will break their dormancy.

The tubers of most varieties of potato emerge from dormancy more rapidly when stored at 22°C than at 10°C.

ARTIFICIAL MEANS OF BREAKING BUD DORMANCY

A very wide range of treatments, especially using various chemicals, has been found to break the dormancy of resting organs. One of the simplest methods of breaking the dormancy of woody plants consists of immersing the shoots in warm water (at 30°–35°C) for 9–12 hours; this treatment is used by florists for "forcing" the flower buds of lilac and *Forsythia*, to get blooms much earlier than normally. Exposure to ether vapour is also effective in removing the dormancy of lilac buds and of lily-of-the-valley rhizomes, the flowers of which can thus be obtained very early in the winter for sale by florists. Among the other substances which have been found very effective in breaking dormancy are thiourea and ethylene chlorhydrin, which will remove the bud dormancy of a wide range of woody plants and are also effective with potato tubers and rhubarb root stocks.

The various substances we have so far mentioned as being effective in breaking dormancy are not known to occur naturally. Considerable interest, however, arises from the discovery that gibberellic acid (GA_3) is known to break the dormancy of the buds of a number of woody plants, including both those with a chilling requirement, e.g. lilac (*Syringa vulgaris*), black currant (*Ribes nigrum*), horse chestnut (*Aesculus hippocastanum*) and those in which bud-break appears to be controlled by daylength, e.g. beech (*Fagus sylvatica*) and *Rhododendron*.

Kinetin has been found to overcome bud dormancy in several species, including grape vine (*Vitis vinifera*) and apple (*Malus*).

HORMONAL CONTROL OF BUD DORMANCY

Since hormones play an important role in many other aspects of growth and development, it is reasonable to examine their possible role in the control of bud dormancy, especially since the growth-promoting substances, gibberellic acid and kinetin, will overcome dormancy in many species.

It is clear that in true dormancy certain metabolic processes are blocked, so that growth cannot take place even though environmental conditions appear favourable. It is possible that the inability of dormant organs to grow is due to a deficiency of growth-promoting hormones. On the other hand, it is now known that growth inhibiting substances are also common in plant tissues, and hence it is possible that dormancy involves the active inhibition of growth by such substances, as was first suggested by Hemberg. He showed that extracts of dormant potato tubers and buds of ash (*Fraxinus excelsior*) contain substances which inhibit the growth of *Avena* coleoptiles, and that treatments, such as exposure to ethylene chlorhydrin, which remove the dormancy of potato tubers, also cause a marked reduction in the levels of the inhibitors. Similarly, it has been shown for a number of woody species that the levels of endogenous inhibitors in resting buds decline during the course of the winter, and this change is correlated with the gradual emergence of the buds from dormancy.

However, it is clear that a correlation of this type does not establish a causal relationship between growth inhibitors and dormancy. Moreover, because certain substances extracted from plant tissues inhibit the growth of coleoptiles it does not necessarily follow that such substances normally inhibit growth in the plants from which they were extracted. It is, therefore, necessary to examine more critically the hypothesis that growth inhibitors regulate dormancy.

Some of the best evidence for the hypothesis comes from a study of the photoperiodic responses in woody plants. The fact that growth of the apical meristems is arrested and resting buds are formed when the leaves are exposed to SD, suggests that the leaves exert an inhibitory effect on the meristems.

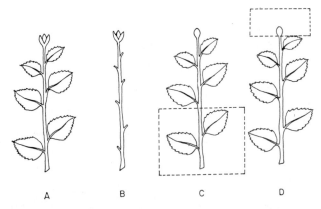

FIG. 102. Effect of leaves on resumption of growth by dormant seedlings of birch, when transferred to long days. (All parts of the plants were exposed to LD,, except those enclosed by broken lines, which were exposed to SD). Buds resume growth when exposed to long days, whether leaves present or absent (A, B). When the buds are exposed to long day and the leaves to short day (C), the buds fail to grow. Exposure of the leaves to long day does not cause the growth of buds maintained under short day (D). Comparison of the seedlings of treatments B and C indicates that short day leaves have an inhibitory effect on bud growth.

There is, however, even more direct evidence of the inhibitory influence of SD leaves on buds. Thus, if seedlings of birch are first rendered dormant by exposing them to SD, and then they are transferred back to LD, then the resting buds will resume growth under the latter conditions provided that either (1) both leaves and buds are exposed to LD, or (2) the plants are defoliated and only the buds are exposed to LD. On the other hand, if the buds are exposed to LD and the leaves to SD, the buds remain dormant (Fig. 102). Evidently some inhibitory influence must be transmitted from the leaves to the buds under SD conditions, and it is reasonable to consider first the possibility that leaves produce growth-inhibitors under SD. It has, indeed, been known for a considerable time that extracts of leaves, including those of several woody species, contain growth inhibiting substances, the region of paper chromatograms containing such substances being referred to as the 'β-inhibitor' zone. Hence, it seemed possible that the inhibitory effect of SD leaves is due to the presence of β-inhibitor.

These latter observations led to attempts to isolate the inhibitors present in the buds and leaves of woody species. It was shown that the leaves of sycamore (*Acer pseudoplatanus*) contain a highly active growth-inhibiting substance, and further work led to the isolation of a crystalline fraction which was shown to be identical with a substance isolated by Addicott and his co-workers from

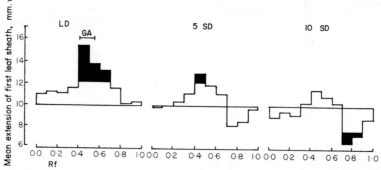

FIG. 103. Hormone changes in shoot apices of black currant (*Ribes nigrum*) in response to short days. Extracts of shoot spices were chromatographed, and the various zones of the chromatograms were assayed for gibberellin and inhibitor changes, using sections of first internode and leaf of dwarf maize. Growth above the water controls indicates gibberellin activity and growth below water controls indicates growth inhibition. *Left:* plants kept under long days; *centre:* plants extracted after five SD cycles; *right:* plants extracted after ten SD cycles. (From H. M. M. El-Antably and P. F. Wareing, unpublished.)

young cotton fruits, which promotes lead abscission in cotton plants, known as *abscisic acid* (ABA). This substance is a sesquiterpenoid (p. 73) having the following structure:

Abscisic acid

The molecule contains an asymmetric carbon atom (C) and therefore exhibits optical isomerism, a property which has been used to show the presence of ABA in a considerable number of other species, including birch, willow and ash. Only the (+) enantimorph has been found to occur naturally.

ABA is a highly active growth inhibitor in a variety of growth tests, as will be described below (p. 237). The availability of synthetic ABA has made it possible to test whether dormancy can be induced in seedlings of woody species by external application of ABA. Experiments of this type have been carried out by applying solutions of ABA to the seedlings, either by dipping a leaf in a tube of the solution, or by feeding it through a lip cut in the stem; in this way it has been found possible to induce the formation of typical resting buds in seedlings of sycamore (*Acer pseudoplatanus*) and black currant, species having "foliar" buds (p. 226). We have seen (p. 226) that foliar bud scales are structures in which the development of the lamina is suppressed, so that the scale represents only the base of a normal leaf. It is not difficult to envisage how the application of a growth inhibitor, such as ABA, may lead to the formation of a bud scale by arresting the development of a leaf primordium at a certain critical stage, so that lamina development is prevented. ABA will also induce the formation of winter resting buds in the duckweed *Lemna polyrhiza* and these can only be induced to resume growth by chilling or by treatment with kinetin.

Earlier observations suggested that higher levels of ABA are present in leaves and apices under SD than under LD, but more recent evidence suggests that the level of ABA is not greatly affected by photoperiod. However, the gibberellin levels decrease markedly on transfer from LD to SD (Fig. 103), and the apparent change in inhibitor levels may be due to variations in the levels of gibberellins which occur in the same region as inhibitors on the chromatograms.

Growth Promoting Hormones and Bud-break

The possibility that endogenous gibberellins may be involved in emergence of buds from dormancy is suggested by the fact that application of exogenous gibberellic acid will induce bud-break in a number of woody species. Moreover, there is a significant increase in gibberellin levels during the later part of the rest period, while the inhibitor levels are declining and the buds are emerging from dormancy. When observations are made on changes in buds exposed to natural fluctuating winter temperatures, it is not possible to determine whether the increased gibberellin levels occur during the period of exposure to chilling, or whether they are found during intervening periods when the temperatures are higher. However, observations carried out on shoots of black currant maintained at a constant chilling temperature of 2°C in a cold room have shown that there is an increase in gibberellin levels during the actual period of chilling (Fig. 104). Similar variations in the levels of endogenous cytokinins during winter chilling of woody plants have been reported. Thus there is an increase in cytokinins in the xylem sap of poplar and other species in response to chilling.

It seems possible, therefore, that the control of bud dormancy may involve an interaction between growth promoters, such as gibberellins and cytokinins, on the one hand, and inhibitors such as ABA, on the other.

Although, as stated above, the levels of endogenous ABA may not vary with photoperiod, since the levels of gibberellins decrease when seedlings of woody plants are transferred from LD to SD, it is still possible that induction of dormancy is controlled by the relative levels of endogenous ABA and gibberellins.

Similarly, in the breaking of bud dormancy by winter chilling, there is considerable variation between species in the extent to which inhibitor levels decrease, and it may be the increase in gibberellin and cytokinin levels which controls emergence from dormancy.

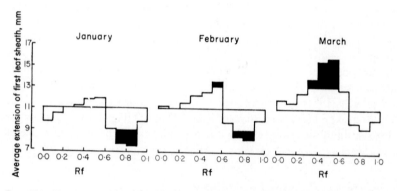

FIG. 104. Changes in gibberellin/inhibitor content of buds of black currant (*Ribes nigrum*) in response to winter chilling. Buds were collected at monthly intervals from plants grown out-of-doors and were extracted and assayed for hormone activity, as described for Fig. 103. (From H. M. M. El-Antably and P. F. Wareing, unpublished.)

However, the state of dormancy may not be regulated directly by the relative levels of gibberellins and ABA present in the tissues at a given time. We shall see later (p. 278) that there is evidence that dormancy may involve a reduction in the capacity of the DNA to support the synthesis of RNA, and that ABA may be involved in this effect. Thus, it is possible that the high levels of ABA present in woody plants under SD conditions result in a blocking of certain important aspects of metabolism. On the other hand, the release from dormancy brought about by winter chilling may involve the "unblocking" of the DNA, possibly by an increase in gibberellins, and the concomitant decline in inhibitor levels may be the result, rather than the cause, of dormancy release. Thus, it is possible that the induction of bud dormancy by SD may be due to high ABA levels, whereas dormancy release may be determined primarily by increased gibberellin and cytokinin levels, rather than by a decrease in ABA.

The Properties and Mode of Action of Abscisic Acid

Since its original isolation from young cotton fruits and leaves of sycamore, ABA has been shown to be present in a wide range of plant species and in a variety of plant tissues and it would appear that it must now be included among the list of hormones involved in the regulation of plant growth and development.

As we have seen, whereas auxins, gibberellins and cytokinins are, in general, growth-promoting hormones, ABA is a growth inhibitor and its mode of action in bringing about bud dormancy can be interpreted as primarily a process of growth inhibition. ABA is also found to be a powerful inhibitor in various growth tests; for example, it causes 50 per cent inhibition of the growth of oat coleoptiles at a concentration of 10^{-6} M, which is well within the range of activity of the growth-promoting hormones. ABA also inhibits gibberellin-stimulated growth in various tests—for example, in maize leaf sections. ABA appears to be much less inhibitory when sprayed on to whole plants, but this apparent lower activity may be due either to poor penetration into the leaf, or to the fact that it is rapidly inactivated in the tissues. Germination of seeds, such as those of lettuce, is inhibited by ABA, and this effect can be overcome by kinetin.

As stated above, ABA was first discovered as the result of studies on its effect in promoting petiole abscission in explants of cotton seedlings, and this effect has now been shown for several other species. In this response its effect is antagonized by IAA.

When leaf discs of a wide range of species are treated with ABA, senescence is greatly accelerated, although it is considerably less effective when applied to intact leaves. In duckweed (*Lemna*) the effect of ABA on leaf senescence can be reversed by the cytokinin, benzyladenine. Another effect of ABA is the inhibition of flowering in several long-day plants, including *Lolium temulentum*.

It is thus apparent that ABA has a considerable variety of physiological effects, and this fact, together with its widespread occurrence, suggests that its physiological role is probably much wider than that of regulating bud dormancy in woody plants, and it may have a general function as a growth regulator, in interaction with the other growth hormones.

The question arises as to the possible mode of action of ABA as a growth inhibitor. Since it appears to antagonize the effect of auxins and gibberellins, and possibly of cytokinins, in various tests, it might be thought that it competes with one of these hormones for a specific enzyme site, in the same way as certain metabolic inhibitors compete with metabolites. If this were so, then it should be possible to overcome the effects of ABA completely by adding a sufficiently high concentration of growth promoting hormone, but in general this is not the case. Another way in which ABA might antagonize the action of a growth-promoting hormone would be to inhibit its biosynthesis, or promote its inactivation, in the plant. Good evidence has been obtained that ABA reduces endogenous gibberellin levels in maize seedlings, but it is not known how general is this effect.

The effect of ABA in promoting leaf senescence suggests that it may inhibit RNA and protein synthesis and this has, indeed, been found to be the case (p. 266). Thus, it is possible that the inhibition of growth by ABA may be brought about by its effect on RNA metabolism, as is believed to be the case with growth-promoting hormones (p. 287).

DORMANCY IN SEEDS

Morphologically, the seed consists of an embryo surrounded by one or more covering structures, of which the most important is the testa, which is usually derived from the integuments of the ovule. Some seeds contain a well-developed endosperm, which lies within the testa and may surround the embryo, or may lie to one side of it. Functionally, the seed is a "propagule" or dispersal unit, i.e. an organ of propagation. In many species the seeds are liberated from the fruit and the isolated seeds become the dispersal units. In other species, however, the fruits may contain a single seed which is retained within the fruit coat (pericarp), the fruit itself being shed as a whole and becoming the dispersal unit. Examples are provided by achenes, nuts, caryopses and so on, the precise definition of which does not concern us here. Although these latter structures are distinguishable morphologically from seeds, they perform the same biological function as seeds, i.e. they are functional propagules and hence it is convenient to refer to all such structures as seeds, although it is not strictly accurate to do so.

Although, as we shall see, dormancy in seeds shows many parallels with that in buds and other organs, the presence of enclosing coats introduces complications not found in buds, and we find several types of seed dormancy, which do not appear to correspond to any form of bud dormancy.

Hard Seed Coats

The seeds of certain families, such as the Leguminosae, Chenopodiaceae, Malvaceae and Geraniaceae, possess testas which are impermeable to water, so that such seeds are liable to lie dormant in the soil for considerable periods before germination occurs. Water uptake by these seeds can be brought about by various treatments, such as abrasion by sand, treatment for short periods with concentrated sulphuric acid, etc., which remove the impermeable outer layer of the testa and permit penetration of water to the embryo. Seedsmen treat clover seed by rotating it in a drum lined with carborundum. Probably under natural conditions the activities of micro-organisms in the soil slowly break down the outer layers of the testa and so render water uptake possible.

Immaturity of the Embryo

In certain seeds the embryo is still immature when the seed is shed and germination cannot occur in such seeds until the embryo has undergone development. This is true of the seeds of wood anemone (*Anemone nemorosa*), lesser celandine (*Ficaria verna*), marsh marigold (*Caltha palustris*), ash (*Fraxinus excelsior*), and other species. In order for this further development to take place, the seeds must be imbibed with water and maintained under favourable temperature conditions. The time required for the embryo to complete its development may vary from about 10 days in *C. palustris* to several months in *F. excelsior*.

After-ripening in Dry Storage

The seeds of many species fail to germinate if sown immediately after harvesting, even though the embryo is fully mature. If they are stored dry at ordinary room temperatures, however, they gradually lose their dormancy and become capable of germinating when provided with suitable conditions (Fig. 105). This effect is called "after-ripening in dry-storage", and is found in several types of cereal, e.g. barley, wheat, oats and rice. The duration of the dormant period may range from a few weeks to several months. Other species showing this type of dormancy include many grasses, black mustard (*Brassica nigra*), evening primrose (*Oenothera* spp.), clover (*Trifolium* spp.), and cultivated varieties of lettuce.

It is not known what causes this type of dormancy nor what are the changes which occur during the storage period which ultimately release the seed from dormancy. It would seem that the processes involved during this period of after-ripening are not of a metabolic nature, since they occur even in the dry seed, when metabolism is at a very low level.

This type of dormancy is of considerable economic importance in cereals. In regions where the weather during the harvest period is liable to be wet, dormancy of the grain is an advantage, since varieties which show such dormancy are less liable to germinate in the ear under wet conditions. From this point of view, dormancy is a desirable economic "character", which is deliberately selected for by plant breeders. On the other hand, for the production of malt from barley, dormancy of the grain is often a major problem, since it may be impossible to germinate the grain for several weeks after harvesting.

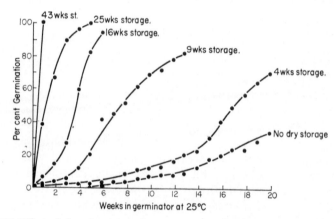

Fig. 105. Effect of after-ripening in dry storage at room temperature on germination rate of seeds of *Impatiens balsamina*. (From W. Crocker, "Growth of Plants", Reinhold, New York, 1949.)

Light-sensitive Seeds

One of the most interesting forms of dormancy, and one which has received intensive study in recent years, is that shown by light-sensitive seeds. In a considerable number of species, exposure to light is necessary for germination, e.g. seeds of tobacco (*Nicotiana* spp.), fox-glove (*Digitalis purpurea*), hairy willow-herb (*Epilobium hirsutum*), purple loosestrife (*Lythrum salicaria*), dock (*Rumex crispus*), and many others. On the other hand, in certain other species germination is *inhibited* by light, although the number of such species is considerably smaller than for light-promoted seeds; among the known light-inhibited seeds are those of love-in-a-mist (*Nigella*), *Nemophila*, *Phacelia* and annual phlox (*Phlox Drummondii*).

Light-sensitive seeds will only respond to light after they have imbibed water. The duration of illumination required by light-promoted seeds is often very short; for example, a high percentage of germination is obtained with lettuce seed exposed to only 1–2 minutes of light, while with seed of purple loosestrife a light-flash of only 0·1 second's duration has a marked effect in stimulating germination. The responses of light-sensitive seeds are strongly affected by temperature and many seeds which are light-requiring at, say, 25°C become capable of germinating in the dark at lower temperatures, e.g. certain light-requiring varieties of lettuce. If certain light-requiring seeds are exposed to daily alternations of temperature, e.g. between 15°C and 25°C, they can be induced to germinate without exposure to light. Other treatments which can replace the light requirement of seeds include treatment with certain inorganic ions, especially nitrate, and certain organic substances, e.g. thiourea.

Many seeds which are light-requiring when freshly harvested, gradually lose their light-requirement during storage and ultimately give full germination

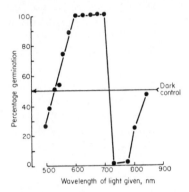

FIG. 106. Effect of red and far-red radiation on germination of lettuce seed. (From L. H. Flint and E. D. McAlister, *Smithsonian Inst. Misc. Collections*, **96**, 1–8, 1937.)

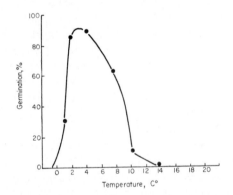

FIG. 107. Effect of chilling temperatures on dormancy of apple seed (germination after 85 days chilling at temperatures shown). (From P. G. de Haas and H. Scharder, *Zeitschrift für Pflanzenzüchtung*, **31**, 457, 1952.)

in complete darkness, e.g. light-sensitive varieties of lettuce. It would seem, therefore, that the changes occurring during after-ripening in dry storage in some way remove the light requirement.

As we have already seen, studies on the responses of light-sensitive varieties of lettuce seed played a key role in the discovery of phytochrome (Chapter 9, p. 179). It was shown that the red region of the spectrum promotes and far-red inhibits the germination of lettuce seed (Fig. 106). If red and far-red are given alternately, then whether germination occurs or not depends upon the nature of the last radiation to which the seeds were exposed. Thus, when phytochrome is converted to the P_{fr} form by red light, it evidently initiates a chain of processes which ultimately result in germination. It has been found that similar red/far-red responses are shown by other species of light-sensitive seeds and it seems probable that phytochrome is universally involved in light-promoted seeds.

Light inhibited seeds have been very much less intensively studied than light-promoted ones, but it now seems probable that the same phytochrome system is involved in both types of seed, and that in the light-inhibited species the effect of far-red is enhanced so that it predominates over the effect of red. Thus, it has been shown that the light-inhibition of *Nemophila* seed is due mainly to the far-red region of the spectrum; red light, on the other hand, seems to have little or no promotive effect on this seed.

Dormancy Removed by Chilling

Gardeners have long known that the seeds of many species will not germinate if sown under warm conditions, but will lie dormant in the soil for long periods; if, however, they are sown out-of-doors in the autumn and exposed to winter conditions they will germinate in the following spring.

This behaviour led to the horticultural practice of "stratifying" the seed, i.e. placing it between layers of sand and leaving it out-of-doors during the winter. Such "stratified" seed is no longer dormant and germinates readily in the spring (Fig. 108). From such observations it is clear that exposure to winter cold is, in some way, necessary to break the dormancy of many seeds.

At one time it was believed that the dormancy of such seeds is due to hard and impermeable coats, and that freezing temperatures are necessary to break the coats. It is now known, however, that freezing temperatures are not required, and that, in fact, temperatures just above freezing (0–5°C) are more effective than lower temperatures (Fig. 107). Moreover, many seeds which have a chilling requirement do not, in fact, have hard coats, e.g. apple, birch.

The range of seeds showing a chilling requirement is very wide (Table 6) and includes both woody and herbaceous plants. In some seeds there is an obligate requirement for chilling, e.g. ash (*Fraxinus excelsior*), whereas in others, e.g. *Pinus* spp., a period of pretreatment at chilling temperature, although not essential, nevertheless increases and hastens subsequent germination. It should be noted that for chilling temperatures to be effective the seeds must be imbibed with water, there being no effect with dry seeds. The minimum period of chilling necessary to remove dormancy varies from species to species, but usually amounts to several weeks. In some species the embryo itself is dormant and can only be induced to germinate with difficulty if it is unchilled, e.g. mountain ash (*Sorbus aucuparia*), whereas in other species the embryo will germinate if the testa is removed and only the intact seed has a chilling requirement, e.g. sycamore (*Acer pseudoplatanus*). Seedlings grown from unchilled

FIG. 108. Effect of winter chilling on dormancy of seeds of *Rhodotypos*. A. Seeds exposed to outdoor temperatures throughout the winter. B. Seeds maintained in a warm greenhouse throughout experiment. (Photograph supplied by late Dr. Lela V. Barton.)

TABLE 6. WOODY SPECIES WITH SEEDS HAVING A CHILLING REQUIREMENT TO OVER-COME DORMANCY

Acer spp. (Maples)	*Malus* spp. (Apple and Crab apple)
Betula spp. (Birches)	*Picea* spp. (Spruces)
Cornus florida (Dogwood)	*Pinus* spp. (Pines)
Corylus avellana (Hazel)	*Prunus* spp. (including Peach)
Crataegus spp. (Hawthorns)	*Rosa* spp. (Roses)
Fagus sylvatica (Beech)	*Sequoiadendron giganteum* (Wellingtonia)
Fraxinus spp. (Ash)	*Tilia* spp. (Lime)
Hamamelis virginiana (Witch hazel)	*Thuja occidentalis* (Western red-cedar)
Juglans nigra (Walnut)	*Vitis* spp. (Grape)
Liriodendron tulipifera (Yellow poplar, Tulip tree)	

embryos frequently show "dwarfism", however, making sluggish growth and having very short internodes. This dwarfism of seedlings can itself be removed by chilling or by treatment with gibberellic acid.

Certain seeds, such as acorns (*Quercus*) and those of *Viburnum*, show "epicotyl dormancy"; such seeds germinate and develop a radicle in the autumn without any prior chilling but development of the epicotyl is dependent upon chilling, i.e. the epicotyl, but not the radicle, shows dormancy.

A few species have "two-year seeds", so called because they do not normally germinate until the second spring after shedding. Certain types of two-year seeds have hard coats, as well as a chilling requirement, e.g. hawthorn (*Crataegus*) and *Cotoneaster*; because of the hard coats, the embryos are prevented from imbibing water as soon as shed, and hence chilling during the first winter is ineffective in removing dormancy. The hard coats are rendered permeable to water during the following summer, however, as a result of the activities of soil micro-organisms. When such imbibed seeds enter the second winter, the dormancy is broken and they become capable of germinating in the following spring.

The cause of the "two-year" behaviour is different in other species. Thus, seeds of lily-of-the-valley (*Convallaria*) and Solomon's seal (*Polygonatum*) require a chilling period to bring about growth of the radicle, but development of the epicotyl does not follow until the seeds have been subjected to a second winter's chilling.

THE ROLE OF THE COATS IN SEED DORMANCY

The seed coats have been found to play an important role in the dormancy of the seed of many species. It has already been mentioned that although the embryos are dormant in some seeds which have a chilling requirement, nevertheless in other species it is only the *intact* seeds which show dormancy, and the isolated embryos will germinate without chilling if the testa is removed. Similarly, certain light-requiring seeds, such as birch and lettuce, will germinate in darkness if the seed coverings are removed or even if they are only slit. Again, certain seeds which show a requirement for after-ripening in dry storage will germinate if the seed coverings are removed; for example, removal of the husks of barley, wheat, oats and rice will permit germination soon after harvesting, whereas when the coats are left intact such seeds normally require several weeks of after-ripening. It is clear, therefore, that the seed coats play an important role in at least three different types of dormancy, and, indeed, in all cases where the embryo itself is not dormant, the dormancy of the intact seed depends upon the presence of coats, which will include the testa, together with the endosperm and pericarp in some seeds. This conclusion raises the question as to the mechanism of these seed coat effects.

It is possible that the seed coats present a physical barrier to gaseous exchange of oxygen and carbon dioxide between the embryo and the external air. It seems unlikely that seed coat effects are due to the accumulation of high internal concentrations of carbon dioxide since germination of lettuce seed is actually *stimulated* by keeping the seeds in high concentrations of this gas. On the other hand, several types of seed show higher oxygen requirements than do actively growing plants of the same species, suggesting that seed coats may present a physical barrier to oxygen uptake. The testas of marrow (*Cucurbita pepo*) seed have been shown to be much less permeable to oxygen than to carbon dioxide. Certain seeds can be induced to germinate either by slitting or removing the coats, or by maintaining the intact seed in a high concentration of oxygen, e.g. in *Betula* and non-after-ripened cereals. Studies on the respiration of germinating pea seeds suggest that anaerobic conditions may occur in the initial stages of germination until the testa is ruptured, when there is a marked increase in oxygen uptake (p. 253). Thus, several kinds of evidence support the view that seed coats may limit the uptake of oxygen.

Interference with oxygen uptake, especially in association with high temperatures, appears to be important in what is known as *secondary dormancy*. Thus, non-dormant seeds of *Xanthium* can be rendered dormant by embedding them in clay (which restricts gaseous exchange) and keeping them at 30°C for several weeks. Similar secondary dormancy phenomena have been shown for a number of other species, including members of the Polygonaceae and Rosaceae, e.g. apple and pear. In all these cases, the secondary dormancy can be overcome by chilling treatment, and it appears that the development of secondary dormancy is the reverse of after-ripening.

Secondary dormancy shows many resemblances to primary dormancy and it has been suggested by Vegis and others that restricted oxygen uptake, in association with high temperature, is the cause of normal dormancy in seeds and, indeed, in buds also. Thus, Vegis has pointed out that the embryos of developing seeds are liable to experience oxygen deficiency because of the surrounding seed coats and maternal tissues, and postulates that under such conditions of partial anaerobiosis normal oxidative breakdown through the tricarboxylic acid cycle and "terminal oxidation", necessary for growth in most species, does not take place. Instead, the products of glycolysis, such as phosphoglyceric acid, cannot undergo normal oxidative breakdown, but become diverted into alternate pathways, leading to the formation of fatty acids and fats, which tend to accumulate in dormant tissues.

The possible importance of oxygen deficiency within the seed as a factor in dormancy is suggested by work on rice seeds, which are dormant immediately after harvesting, but gradually emerge from dormancy during dry storage. The dormant seeds of rice can be induced to germinate by removing the husks, thus indicating the importance of coat effects in these seeds. Storage in oxygen greatly reduces the dormancy period, suggesting that some oxidation reaction may be involved in the after-ripening processes during dry storage. However, Roberts tested the effects of various respiratory inhibitors (including inhibitors of terminal oxidation, Krebs cycle and glycolysis) and obtained the unexpected result that they *stimulated* germination of dormant rice seeds. He suggested that it is necessary for some oxidation reaction to proceed before germination can take place, and that this reaction is in competition with respiratory processes involving glycolysis, the Krebs cycle and the terminal oxidase system for the low levels of oxygen present in the seeds; hence the inhibition of these respiratory processes by various substances will release greater amounts of oxygen for the other oxidation reaction, which he suggests may involve the "pentose phosphate pathway" of carbohydrate metabolism.

There is also some evidence that the effect of seed coats may be due to mechanical resistance to the growth of the radicle. Thus, several types of dormant seed will germinate if the seed coat is removed in the radicle region, but if seeds so treated are placed in a high osmotic concentration of 0·3 M mannitol (which will reduce the ability of the seeds to take up water and hence replace the mechanical effect of the seed coat) their germination is inhibited. However, the osmotic effect of the mannitol solution can be overcome by treatments, such as exposure to light or treatment with gibberellic acid, which will stimulate germination of the intact seed. It is concluded, therefore, that the normal effect of the seed coat is a mechanical one, to overcome which the radicle needs to develop a sufficient turgor. Whether this mechanical effect of seed coats is important in many species is not yet clear. Nevertheless, whatever the effect of the coats, it is clear that they play a very important role in many forms of seed dormancy.

SIMILARITIES BETWEEN SEED DORMANCY AND BUD DORMANCY

Inhibition of germination by the seed coats cannot be important in seeds which show embryo dormancy, where even the naked embryos are dormant. Hence we must seek some other cause of dormancy in such cases. Now the dormancy of seeds showing a light- or chilling-requirement shows certain features in common with that of buds and other organs, which may be summarized briefly as follows:

1. Chilling for several weeks at 0–5°C is effective in breaking the dormancy of buds, rhizomes, corms and many types of seed.
2. Certain substances will break the dormancy of several kinds of organ; thus, thiourea and gibberellic acid will remove the dormancy of tree buds, potato tubers and several types of seed.
3. Certain tree buds and certain seeds may be induced to grow by exposure to long days, whereas short days are ineffective.

The close parallel between dormancy in buds and in seeds is particularly clear in instances where the buds and seeds of a single species are compared. For example, in birch (*Betula pubescens*) the dormancy of both the seeds and buds can be removed by chilling, by exposure to long days or by gibberellic acid. This parallel between seed and bud dormancy in a single species strongly suggests that the cause of dormancy is the same in both organs.

Now, bud dormancy is apparently not due to interference with gaseous exchange by the bud scales since (1) many dormant buds are not tightly enclosed by the scales, and (2) removal of the bud scales does not usually cause resumption of apical activity. Moreover, interference with gaseous exchange cannot be important in the *induction* of dormancy in buds, since until the buds are actually formed there can be no question of interference with oxygen uptake by the bud scales. On the other hand, we have seen that in many woody species resting buds are formed under short days, and that the response is determined by the daylength conditions to which the *leaves* are exposed. In view of the evidence for the role of hormones in the control of bud dormancy it is pertinent to examine their possible importance in some forms of seed dormancy.

GERMINATION INHIBITORS AND SEED DORMANCY

The possible importance of germination-inhibiting substances was first suggested by Molisch in 1922. He and his co-workers studied the problem of why the seeds of succulent fruits, such as those of tomato and cucumber, do not normally germinate in the fruit, but they will readily germinate if they are removed, washed and sown on moist filter paper. They showed that the sap of several fruits will inhibit the germination of seeds of the same species when the latter are planted on filter paper containing the sap. It was shown that the inhibitory effect of tomato juice is not primarily due to its acid pH or high osmotic pressure, and it thus appears to be due to the presence of substances which inhibit germination by virtue of their chemical properties. Tomato juice has been shown to contain a number of phenolic substances, including caffeic and ferulic acids, but it is not clear whether the inhibition is due to these substances.

There is also evidence that germination inhibitors may be important in non-succulent fruits. For example, the "seed balls" (actually clusters of fruits) of beet show difficulties in germination, and it has been shown that the outer coat (i.e. pericarp) contains powerful germination inhibitors. There seems to be good evidence that these inhibitors are partly responsible for the delayed germination of beet seed, since the latter will not germinate if they are allowed to take up a limited amount of water from moist filter paper, but they will germinate if they are first washed with a considerable quantity of water, which leaches out the inhibitors from the outer coats.

There is thus evidence that germination inhibitors may prevent premature germination in succulent fruits and may cause delayed germination in some non-succulent fruits. From this situation it is but a short step to envisage that inhibitors present in the testa, endosperm or embryo itself may cause delayed germination and hence give rise to seed dormancy.

The possible role of inhibitors in the dormancy of seeds showing a chilling requirement has been studied in several species. Inhibitors have been shown to be present in the embryos and seed coverings, but by contrast with the situation in buds, the levels of these inhibitors do not appear to decline appreciably in response to chilling. On the other hand, leaching of dormant embryos, which presumably removes or dilutes the inhibitors, leads to germination of unchilled embryos in ash (*Fraxinus excelsior*) and several species of *Rosa*. In *Rosa* the pericarp, which has a strongly inhibitory effect upon the embryo, has been shown to contain relatively high concentrations of abscisic acid, and it is likely that the inhibitory effect of the pericarp is partly due to the presence of this substance. ABA is present also in seeds of ash and peach.

There is evidence that inhibitors may be involved in the dormancy of certain light-requiring seeds. For example, the intact achenes ("seeds") of birch are light-requiring, but if the embryos themselves are removed from the achenes they will germinate equally well in light or dark. Thus, the coats (testa, endosperm and pericarp) which normally enclose the embryo appear to inhibit its germination, so that it requires light to overcome the inhibitory effect. Now it has been shown that the endosperm and pericarp of birch seeds contain a germination inhibitor and if the embryos are placed on filter paper soaked in the inhibitor they no longer will germinate in the dark, but will do so in the light. Thus, the inhibitory effect of the pericarp and endosperm of the intact seed appears to be due partly to the inhibitory substances which they contain. On the other hand, simply slitting the "seed" coats (pericarp, endosperm and testa) of birch achenes leads to considerably increased germination in the dark, suggesting that part of the inhibitory effect of the coats is due to interference with gaseous exchange. Thus, the effect of the coats in this species appears to involve both inhibitory substances and interference with gaseous exchange.

As in buds, the dormancy responses of seeds with a chilling or light requirement appear to involve an interaction between inhibitors and germination promoters, especially gibberellins. Application of GA_3 will stimulate the germination of certain light-requiring seeds, e.g. lettuce, birch, and also of seeds which normally require chilling, e.g. beech, hazel and oak. Thus, GA_3 can substitute for both light or chilling in the seed of various species. It is possible, therefore, that one effect of chilling is to increase the level of endogenous gibberellins, and some evidence for this effect has been obtained for seeds of ash and hazel. In ash, the endogenous germination promoter, which increases during chilling, will overcome the effects of the natural inhibitor which is present in the endosperm and pericarp. In hazel, however, there is only a small increase in gibberellin levels during the actual chilling period, and it is only when the seeds are transferred to warm conditions that the gibberellin levels increase markedly. Thus, in this species it would appear that the increase in gibberellins is the result of certain other changes brought about by chilling. In seeds of sugar maple (*Acer saccharum*) increases in levels of both gibberellins and cytokinins have been found during chilling.

Interaction between Coat Effects and Inhibitors

It would appear that "coat dormancy" (i.e. dormancy imposed by the coats) and embryo dormancy in seeds involve different mechanisms, the former resulting from interference with gaseous exchange and the latter from the presence of growth inhibitors. At first sight these two causes of dormancy would appear to have quite distinct and separate modes of action, but there is some evidence that there may be an interaction between coat effects and those of inhibitors. We have already seen that dormancy in birch seed appears to involve both factors. Again, in certain species, such as sycamore (*Acer pseudoplatanus*) the seeds show embryo dormancy shortly after they are shed but after a period of storage the isolated embryos become non-dormant although the intact seed continues to remain dormant until chilled. Thus, we see a change from one type of dormancy to the other, within a single seed, indicating that there may not be a sharp distinction between them.

The close connection between coat effects and inhibitors is also illustrated by *Xanthium pennsylvanicum*, the fruit of which contains two seeds, one of which is dormant, and the other non-dormant. There is a marked coat effect in the dormant seed, since removal of the coat permits germination. The intact seed will germinate in an atmosphere of pure oxygen, which was formerly taken to indicate that interference with oxygen uptake is important. On the other hand, the embryo contains a powerful germination inhibitor, which is leached out when the coat is removed. If care is taken to avoid the leaching-out of inhibitors, the isolated embryo is found to be dormant and to show a high oxygen requirement for germination. Thus, the high oxygen requirement of intact seeds appears to be due to the presence of the inhibitors in the embryo, and one of the effects of the coat is to prevent leaching of the inhibitor from the intact seed.

In charlock (*Sinapis arvensis*), on the other hand, the relative impermeability of the coat to oxygen appears to lead to the formation of an inhibitor within the embryo.

It is clear that the interaction between coat effects and inhibitors is complex and requires further study.

THE LONGEVITY OF SEEDS

The period during which seeds retain their viability varies greatly between species. The seeds of certain species remain viable for only short periods if kept in air at ordinary atmospheric conditions of humidity and temperature. Thus, seeds of willow (*Salix*) are viable for only a few days and must be sown very shortly after attaining maturity. The seeds of poplar and elm retain viability for a few weeks, and those of oak, beech and hazel for a few months, when stored under cool, moist conditions.

The seeds of most species, however, remain viable for considerable periods, generally for at least a year and frequently for much longer. Various types of experiments have been carried out to determine the longevity of such seeds. Thus, various workers have tested the viability of very old seeds from herbaria, and Becquerel found appreciable germination of seeds of various members of the Leguminosae ranging from 100 to 200 years old. Other families characterized by long-lived seeds are the Euphorbiaceae, Malvaceae, Convolvulaceae, Solanaceae, Labiatae and Compositae.

Several long-term experiments have been set up in the United States to determine the longevity of seeds. In 1902, the U.S. Department of Agriculture set up an experiment in which seed of 107 species were placed in sterile soil in pots, which were then buried outside at different depths. After 20 years, some seeds of 51 of the species were still viable, but the seeds of most cultivated plants tested were dead. After 39 years, low germination was shown by 20 species and high germination by a further 16 species.

A number of field observations also support the finding that seeds may retain their viability for long periods in the soil, under natural conditions. Thus, there is a well-authenticated case in which viable seeds of arable weeds were found in the soils of forests which had been planted on farm land some 20–46 years previously and had presumably lain dormant in the soil for that time. Other similar examples are known. Thus, living seeds of the Indian water lily, *Nelumbium nucifera*, were found in the bed of a former lake in Manchuria, which was estimated to have been drained at least 120 years, and more probably 200 years, previously. Seed from herbarium specimens of this species which were 237 years old have also germinated. The seeds of *Nelumbium* have very thick coats and are impervious to water, so that the embryo is not imbibed with water until the coat is rendered permeable. Similarly, the hard-coated seeds of the Leguminosae will survive in the soil without taking up water until the coats have been eroded by the activities of soil micro-organisms. Such seeds may therefore lie for long periods in the soil in the *unimbibed* condition.

Many other types of seed which survive in the soil for long periods do not have impermeable coats, however, and hence they must survive in a moist condition. This fact raises the problem as to why it is that these seeds do not germinate in the soil, where they appear to have adequate conditions of moisture and temperature for germination. This problem has not been satisfactorily solved, but it has been suggested that high carbon dioxide concentrations in the soil render the seeds dormant. However, recent studies have shown that a very high proportion of buried weeds have a light-requirement for germination. It is of interest that among the buried seeds found to be light-requiring are those of certain species which do not normally show a light-requirement, and hence it would appear that burial in some way leads to the development of a light requirement. Where such light-requiring seeds lie buried in the soil they will remain dormant until ploughing or some other disturbance brings them to the surface, when they rapidly germinate. Seeds showing this behaviour include those of *Digitalis purpurea, Juncus* spp., *Polygonum* spp., *Veronica persica, Spergula arvensis* and *Hieracium* spp.

The ultimate cause of the loss of viability of seeds is not understood, but it is known that seedlings from old seeds show various cytological abnormalities such as chromosome breakage, disturbance of the mitotic spindle, etc., and physiological abnormalities such as chlorophyll deficiency. It is likely, therefore, that the loss of viability is due to slow chemical changes such as the denaturation of proteins.

GERMINATION

The Conditions for Germination

Since the tissues of the ripe seed are in a highly dehydrated condition, it is not surprising that water supply is frequently a limiting factor controlling the germination of seeds. Most seeds take up a relatively large amount of water when planted in a moist medium, and this is initially an imbibitional process, in which various substances present in the seed, especially proteins and starch, are involved. The imbibitional forces involved are enormous and certain seeds can take up considerable quantities of water from relatively dry soil.

Temperature is a second factor which plays an important role in controlling germination. The minimum temperature at which germination can occur varies considerably from one species to another, and the seeds of some species, such as beech, will germinate at temperatures only a little above freezing, whereas the seeds of tropical and sub-tropical species have much higher temperature requirements. Whereas most species will germinate under constant temperature conditions, other species require a daily alternation in temperature. For example, the seed of the dock, *Rumex obtusifolius*, germinates best when subjected to daily temperature alternations of 15° and 30°C. Alternating temperatures are also required by seeds of evening primrose (*Oenothera biennis*), Yorkshire fog grass (*Holcus lanatus*) and celery (*Apium graveolens*). Presumably these require-ments are usually met by the normal variation between day and night tempera-tures under natural conditions. Little is known of the physiological basis of this phenomenon.

Although it is well known that seeds need conditions of adequate aeration for germination, there is little precise information on the oxygen requirements of seeds. It is clear, however, that the oxygen requirements of the intact seed will depend not only upon the metabolic demands of the embryo, but also on the permeability of the enclosing testa or other seed coats. As we have seen, these seed coat effects appear to play an important role in the dormancy of many species. Whereas the oxygen content of the air is far above that needed for normal growth of plants, it is not far above that required for the germination of many seeds, no doubt due to the physical barrier to oxygen uptake presented by the seed coats.

There are marked differences between species in the ability of their seeds to germinate under water, no doubt because the partial pressure of the oxygen dissolved in water is considerably less than in air. Other seeds will germinate well under water; these latter include both aquatic species, and certain land species, such as rice.

The Process of Germination

In non-dormant seeds, active metabolism evidently commences soon after they are placed under conditions favourable for germination. The question arises as to what are the first stages in the complex overall process known as ger-mination. Usually, of course, we take the emergence of the radicle as the primary criterion of germination, i.e. for most practical purposes the initiation of growth is taken as the first detectable sign of germination. It appears that the initial elongation of the radicle involves cell extension rather than cell-division, but cell division starts very early in the growth of the radicle. It is probable that the commencement of radicle growth is preceded by a number of preliminary processes, but little is known regarding the nature of these processes.

In the dry condition of the resting seed, metabolism must obviously be at an extremely low level on account of lack of water, but full metabolic activity does not develop immediately water is imbibed, even in non-dormant seeds.

When a non-dormant seed is planted under conditions favourable for germination there is a rapid increase in the respiration rate, which can be detected 2–4 hours after soaking in the case of peas. After this initial rise the respiration rate in peas reaches a steady value which is maintained for several hours. At about the time that the testa is broken by the radicle there is a further rapid rise in respiration rate, suggesting that in the initial phases of germination gaseous exchange is limited by the testa. By contrast, in barley and wheat there is a fairly uniform rise of respiration rate during germination.

The changes in carbon dioxide output (Q_{CO_2}), of oxygen uptake (Q_{O_2}) and of the respiratory quotient (R.Q.) during germination provide an indication of the type of respiration occurring (i.e. whether aerobic or anaerobic) and of the nature of the respiratory substrate, i.e. whether carbohydrate, fat or protein. During the early stages of germination of peas respiration appears to be predominantly anaerobic, owing to the restriction of oxygen uptake by the testa, and ethanol may accumulate in the tissues. The enzymes of glycolysis have been shown to be present in pea seeds.

The enzymes of the tricarboxylic acid (TCA) cycle by which aerobic respiration occurs, are located in the mitochondria; it appears that the mitochondria in dry seeds are not fully active and are incapable of carrying out oxidative phosphorylation, but their activity increases during the later stages of germination. Also, it has been found that the electron transport system involved in terminal oxidation is not active in dry seeds. On the other hand, the pentose phosphate pathway, which provides an alternative mechanism for the aerobic respiration of carbohydrates, is active in bean seeds.

Although many enzymes are present in dry seeds, other enzymes are absent or present in an inactive form, and their activity only appears as germination progresses; examples of these are provided by several of the amylases, lipases and proteases responsible for the breakdown of reserve materials during germination. It has clearly been demonstrated that certain enzymes are synthesized *de novo* during germination. An excellent example is provided by the enzyme α-amylase, which is not present in the dry barley grain, but which appears during germination (p. 130). It is apparently secreted by the aleurone layer and brings about the breakdown of starch in the endosperm. Normally the presence of the embryo is necessary for the appearance of α-amylase, suggesting that the production of the enzyme depends upon the supply of some substance from the embryo. It has subsequently been shown that the synthesis of α-amylase can take place in barley grains from which the embryo has been removed, if they are supplied with gibberellic acid, suggesting that the embryo normally supplies a natural gibberellin which initiates α-amylase synthesis in the aleurone layer. In this way enzyme synthesis is regulated, so that it does not take place until the embryo commences growth. It is now known that the synthesis of other enzymes, including RNA-ase and proteolytic enzymes, may be stimulated in barley grains by gibberellic acid.

Studies have been carried out on the RNA changes occurring in seeds during germination. It is found that there is active RNA synthesis during germination, affecting all fractions of RNA, and, in particular, there is marked increase in the monoribosome and polyribosome fractions.* During the first 30 minutes of imbibition of water by wheat embryos there is a rapid formation of functional polyribosomes, with a corresponding increase in protein synthesis. Preparations of ribosomes from dry wheat embryos show little capacity to incorporate ^{14}C-leucine into protein, whereas preparations from imbibed embryos do so. However, if polyuridylic acid is supplied (as a synthetic messenger RNA), the ribosomes of dry seeds are found to be as active as those from imbibed seed. Thus, the inability of the ribosomes from dry seeds to carry out protein synthesis apparently does not lie in any deficiency in the ribosomes themselves, but is apparently due to the unavailability of m-RNA. However, there is evidence that m-RNA is present in dry seeds, but is in some way spatially separated from the ribosomes, and it has been suggested that this m-RNA becomes released during the early stages of germination and is then available to combine with ribosomes and hence carry out protein synthesis.

FURTHER READING

General

MAYER, A. and A. POLJAKOFF-MAYBER. *The Germination of Seeds*, Pergamon Press, Oxford, 1963.

More Advanced Reading

Articles by L. V. BARTON, A. LANG, P. STOKES, M. EVENARI and P. F. WAREING in *Encycl. Plant Physiol.* **15**(1), 1965.

KOLLER, D. A. M. MAYER and A. POLJAKOFF-MAYBER. Seed germination, *Ann. Rev. Plant Physiol.* **13**, 437, 1962.

SAMISH, R. M. Dormancy in woody plants. *Ann. Rev. Plant Physiol.* **5**, 183, 1954.

VEGIS, A. Dormancy in higher plants. *Ann. Rev. Plant Physiol.* **15**, 185, 1964.

VEGIS, A. Climatic control of germination, bud break and dormancy. In *Environmental Control of Plant Growth* (Ed. L. T. Evans), Academic Press, New York, 1963.

WAREING, P. F. Photoperiodism in woody plants. *Ann. Rev. Plant Physiol.* **7**, 191, 1956.

WAREING, P. F. Germination and dormancy, in *Physiology of Plant Growth and Development* (Ed. M. B. Wilkins), McGraw-Hill, London, 1969.

---

* See p. 274 for a summary of the process of protein synthesis.

CHAPTER 12

Senescence

In common with all multicellular organisms, higher plants are mortal and the life of the individual plant is ultimately terminated by death. Before the death of the whole plant has occurred, however, it is likely that there will have been earlier death of a number of its organs and tissues. As a result of the activity of the apical meristem, the upper part of the shoot shows a prolonged embryonic condition, while at the same time senescence and death is occurring in the older lateral organs, notably the leaves, flower parts and fruits. In many plants the death of organs has a pronounced seasonal character and there is a regular annual loss of part of the shoot system. In trees, this annual loss is mainly confined to the lateral organs mentioned, but in some herbaceous perennials, e.g. dock (*Rumex*), nettle (*Urtica*), bracken (*Pteridium aquilinum*), the whole above-ground part of the shoot may die each year. In annual species, the whole plant, except the seeds, dies after flowering and fruiting. Thus, the death of plant parts is a regular feature of the annual cycle of growth and is much more common in plants than in animals.

Preceding the death of an organ or of the whole plant, certain deteriorative changes occur in the tissues which are referred to as *senescence*, and which will be described in detail later. It is clear that we may distinguish between *organ senescence* and *whole plant senescence*. In most plants each leaf has only a limited life span so that as the shoot continues to grow in height, the older leaves at the base tend to senesce and die progressively (Fig. 109). This pattern of senescence has been described as *sequential senescence*, and it must be distinguished from the *simultaneous* or *synchronous senescence* of leaves of temperate deciduous trees, which is so conspicuous in the 'fall'. *Fruit senescence* is seen during ripening of both succulent and non-succulent fruits. The ripening of succulent fruits is a complex process which ultimately terminates in the senescence and decay of the tissues.

Before considering plant senescence it is necessary to distinguish between *monocarpic* species, which flower and fruit only once and then die, from *poly-*

carpic plants, which flower and fruit repeatedly. Monocarpic species include all annual and biennial plants and also a certain number of perennial plants which grow vegetatively for a number of years and then suddenly flower, fruit and die. Examples of this latter type of plant are seen in the "Century Plant" (*Agave*) and bamboo, both of which may grow vegetatively for many years, before the single reproductive phase occurs. Thus, in monocarpic species, death of the whole plant is closely connected with reproduction and is evidently genetically determined to occur at this stage in the life cycle. By contrast, in polycarpic

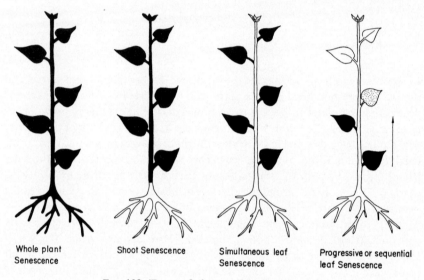

Whole plant Senescence   Shoot Senescence   Simultaneous leaf Senescence   Progressive or sequential leaf Senescence

Fig. 109. Types of plant- and leaf-senescence.

species, which include both herbaceous perennials and woody plants, death of the whole plant is not normally associated with reproduction and there is great variation among the different individuals of a given species with respect to the length of the life span. Thus, in dicotyledonous woody plants, which are all polycarpic, the individual tree normally lives for many years, and there is no universal life span characteristic of the species. Indeed, apart from accident and disease, there would seem no reason why a tree should not live indefinitely, although there would no doubt be mechanical problems when the branches become too heavy to support themselves. In monocarpic plants, on the other hand, death is destined to occur at a given point in the life cycle, and we may speak of *programmed death* in such plants.

What determines the life span of an individual cell, an organ or a whole plant? It might be suggested that a given cell has only a limited life span, which is determined by factors inherent in the cell itself. The loss of viability of seeds appears to be due to chromosome breakage within individual cells (p. 251), but it cannot be assumed that changes occurring in the dehydrated tissues of seeds in dry storage occur also in actively metabolizing cells and are a cause of senescence in the later stages of the plant cycle, although it has been suggested that chromosomal changes are important in the ageing and senescence of animal cells.

In some types of tissue, differentiation involves the early death of certain cells, such as those of vessel elements in the xylem, whereas neighbouring parenchymatous cells may remain living for many years. The changes occurring in the protoplast during the differentiation of a vessel element may correspond closely to those occurring later in the constituent cells of a senescent organ, such as a leaf. However, the process of vacuolation and enlargement does not necessarily involve degenerative changes, since parenchymatous cells may live for many years, as in those of the pith and rays of some woody plants. Thus, it would seem that the maximum potential life of many types of differentiated plant cell is seldom reached in herbaceous plants, and that senescence and death do not occur on account of factors intrinsic in the individual cells, but because of conditions prevailing within the organ or organism as a whole. For example, sequential leaf senescence seems to be caused by competition between mature leaves and the growing regions of the shoot, and if a leaf is removed and allowed to form roots on the petiole it may live very much longer than if it had remained attached to the parent plant (p. 264). Thus the rate of senescence of plant organs is often under control of the whole plant and is not simply determined by intrinsic characteristics of the cells of that organ. However, it would appear that certain organs show inherent senescence processes, which are not under the control of the whole plant; thus, flowers and fruits undergo senescence whether they are allowed to remain on the parent plant or not.

In addition to the various "internal" factors of the plant which are involved in the regulation of senescence, a number of external factors may affect the rate of senescence, including drought, mineral nutrition, light intensity and daylength, and disease. We shall mainly be concerned with a consideration of the internal factors, but the importance of environmental factors must also be borne in mind.

THE BIOLOGICAL SIGNIFICANCE OF SENESCENCE

If we are correct in regarding senescence in monocarpic plants as "genetically programmed", then this implies that the process of senescence has arisen as a result of natural selection and that it has certain biological advantages to the species. What are these advantages? Why, for example, should it be any advantage to an annual species for the whole shoot to senesce during the development and ripening of the fruit—why should the leaves and other parts of the shoot not remain green, as they do during the ripening of the fruit in many polycarpic plants? The answer to this question probably lies in the fact that during the ripening of annual plants there is considerable breakdown of proteins to amino acids, which are exported from the leaves to the developing seeds and re-utilized there as reserve material. For example, in the oat plant a large proportion of the nitrogen in the senescing leaves is transported to the fruits and accumulated there. Thus, recovery of nutrients from senescing organs constitutes a valuable saving to the rest of the plant. Similar export of reserve material occurs during the simultaneous senescence of the leaves of trees in the fall, the exported material being stored in the stem in this case. However, another advantage of leaf fall in deciduous trees probably lies in the resulting reduced rate of transpiration, which is probably essential for survival in climates in which the soil is frozen during the winter (p. 224), while at the same time the return of leaf material and its breakdown in the surface litter releases mineral nutrients to the soil which are available for re-utilization.

Sequential senescence of the basal leaves of a shoot not only releases reserves of nitrogenous and other substances for the young growing leaves, but may also bring about a saving of carbohydrates and other photosynthates where the basal leaves are heavily shaded and hence might actually become "parasitic" upon the plant by importing photosynthates from other parts which they would only consume in useless respiration. Thus, it is not difficult to see how senescence, as an active "programmed" process may have several advantages.

SEQUENTIAL LEAF SENESCENCE

Our present knowledge of the physiology and biochemistry of senescence is derived mainly from studies on leaves, particularly attached leaves undergoing sequential senescence and detached leaves maintained in the dark.

The first visible sign of senescence is yellowing of the leaf, due to the breakdown of chlorophyll which renders visible the other leaf pigments, particularly the xanthophylls and carotenoids. A study of the fine structure of senescing leaves shows that there is progressive degeneration of the membrane structure of the grana of the chloroplasts, accompanied by the appearance of dense globules of lipid material (probably formed from the broken-down membranes) in which the carotenoid pigments dissolve. Other early changes involve the degeneration of the endoplasmic reticulum and the gradual disappearance of the ribosomes. The mitochondria retain their structure during the early stages of senescence, but later they undergo degeneration. The cells of the fully senescent bean leaf still retain an intact plasmalemma but the tonoplast disappears, and the structure of the cytoplasm and nucleus is almost completely lost, leaving the chloroplasts represented by vesicles containing lipid globules.

These structural changes in the cells of the senescing leaf are accompanied by changes in composition and metabolic activity. The protein content of the leaf declines progressively, as a result of the breakdown of proteins to amino acids and amides (Fig. 110). There is also a progressive decline in the RNA content of the leaf, with a particularly marked fall in ribosomal RNA (Fig. 111).

These degenerative changes are reflected in the rate of photosynthesis and respiration of the leaf. In *Perilla*, the rate of photosynthesis declines gradually from the time of full leaf expansion, accelerating during the later stages of senescence (Fig. 112). In this species the trend in photosynthetic rate follows closely the level of the soluble protein fraction of the chloroplasts known as "Fraction I" containing the enzyme, ribulose-1,5-diphosphate carboxylase, which catalyzes the process of carbon dioxide fixation. The trend in respiration

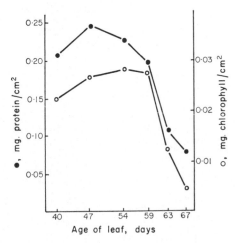

Fig. 110. Changes in protein and chlorophyll content of attached leaves of *Perilla frutescens* from expansion to abscission. (From H. W. Woolhouse, *Symp. Soc. Exp. Biol.* **21,** 179, 1967.)

rate seems to vary according to the species, but in some the rate appears to remain fairly constant until the final stages of senescence, when there is a sharp rise in respiration rate, corresponding to the "climacteric" observed during fruit ripening and senescence.

The question arises as to what initiates and controls the degradative changes occurring during leaf senescence. The observation that, in some species at least, the respiration rate remains constant during the early stages of senescence suggests that it is not changes in respiratory metabolism which cause senescence. On the other hand, we have seen that a constant concomitant of senescence is a marked decline in the protein and RNA contents of leaves, and close attention has been paid to these changes as possibly indicating the "key" processes in senescence. Now, it has been shown that a certain proportion of the leaf protein undergoes continuous "turnover", i.e. the protein is being continuously synthesized and broken down, so that the overall rate of change in protein content represents the net differences in the rates of these two processes. Where there

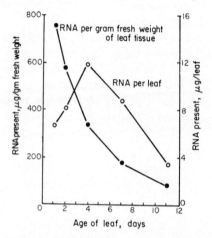

FIG. 111. Changes in ribonucleic acid (RNA) content of attached leaves of pea (*Pisum sativum*) from completion of leaf expansion. (From R. M. Smillie and G. Krotov, *Canad. J. Bot.* **39**, 891, 1961.)

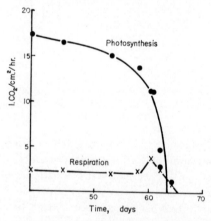

FIG. 112. Changes in rates of photosynthesis and respiration of attached leaves of *Perilla frutescens* from completion of expansion to abscission. (From H. W. Woolhouse, as for Fig. 110.)

is such a continuous turnover, a decline in protein content may reflect a fall in the rate of synthesis or a rise in the rate of breakdown, or both.

The breakdown of protein is brought about by proteolytic enzymes. Studies on changes in proteolytic enzyme activity have given no indication of increased activity during leaf senescence, and hence it would appear that the decline in protein content is primarily due to a reduced rate of synthesis. One possibility that has been suggested is that the senescent leaf retains its full capacity for protein synthesis and that the rate of synthesis is limited by lack of amino acids in the leaf. In a healthy, green leaf the amino acids released by protein breakdown are re-utilized in further protein synthesis. But it has been suggested that in a senescent leaf amino acids are exported to other parts of the plant so rapidly that there is no "pool" of free amino acids available for protein synthesis, so that there is a decline in protein content (Fig. 113). It is, in fact, found that there is no appreciable accumulation of free amino acids in senescent attached leaves, no doubt due to continuous export. However, it is then necessary to explain why the supply of amino acids should be limiting in a senescent leaf and not in a normal one.

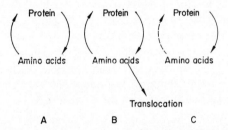

Protein   Protein   Protein

Amino acids   Amino acids   Amino acids

Translocation

A   B   C

Fig. 113. Diagram to illustrate (A) turnover of leaf protein, (B) the translocation hypothesis, and (C) the hypothesis of a defect in protein synthesis. (From E. W. Simon, *Symp. Soc. Exp. Biol.* **21**, 215, 1967.)

Now, there is evidence that the sequential senescence of leaves arises from competition for metabolites and nutrients between old leaves at the base of the stem and young growing leaves in the apical regions. This conclusion is indicated by the fact that if a plant is decapitated and the axillary shoots are removed then the senescence of the older remaining leaves is greatly retarded and indeed leaves which are already showing signs of senescence may undergo a recovery and become green again, even if they were previously showing yellowing. The uppermost leaf of a decapitated plant may also undergo considerable growth and become abnormally large. Thus, it would appear that sequential senescence is a "correlative phenonemon", and shows resemblances to apical dominance (p. 122). Sequential senescence is more pronounced under conditions of mineral nutrient deficiency—for example, plants grown in too small pots frequently show marked senescence of the basal leaves, under which conditions there is presumably severe competition between young and old leaves for available nutrients, and in this competition the young leaves are evidently at an advantage. However, the senescence of the older leaves can be retarded or reversed by the application of nitrogenous nutrients, such as ammonium nitrate.

Thus, it is possible that the competition between young and old leaves results in a rapid rate of transport of amino acids from old to growing leaves and hence to a lower pool of these metabolites being available for protein synthesis within the old leaves. It is an essential part of this hypothesis that there should be active protein turnover, but in fully-expanded *Perilla* leaves the "Fraction I" protein is found to have zero turnover rate, i.e. there is no continuous synthesis and breakdown, and yet during senescence Fraction I declines more rapidly than a second fraction "Fraction 2", which does show active turnover.

Another postulate of the hypothesis we are considering is that the capacity for protein synthesis remains relatively unimpaired during leaf senescence. A convenient method for measuring the rate of protein synthesis is to determine the rate of incorporation of radioactive amino acids, such as ^{14}C-leucine, into protein. Similarly, the rate of RNA synthesis can be followed by the rate of incorporation of an RNA precursor, such as ^{14}C-adenine. Studies of this type have shown that the capacity of tobacco leaves to incorporate ^{14}C-leucine and ^{14}C-adenine declines during senescence, although quite yellow leaves retain some capacity to synthesize certain enzymes, such as peroxidase and ribonuclease (which brings about the breakdown of RNA). It might be argued, however, that the decline in capacity for protein synthesis is the *result*, rather than the cause of senescence.

Thus we see that some evidence supports the hypothesis that the decline in protein content of senescent leaves is due to limiting levels of amino acids, rather than to a reduced capacity for protein synthesis, but at the present time the evidence is not conclusive. We shall now consider an alternative approach to the problem of leaf senescence, in which the possible hormonal control of RNA synthesis is stressed.

SENESCENCE OF DETACHED LEAVES

So far, we have considered the natural senescence of leaves still attached to the plant, but it has been known for many years that when a green leaf is detached from its parent plant it rapidly deteriorates and shows signs of accelerated senescence.

As with attached leaves, the visible signs of senescence are accompanied by a decrease in the protein and RNA contents of the leaf. Protein breakdown commences remarkably soon after the leaf has been detached; for example, protein breakdown is detectable 6 hours after excision of barley leaves. The breakdown of protein commences at the same rate whether the leaves are kept in the light or the dark, but the rate of breakdown later decreases in the light, whereas it continues at a high level in the dark (Fig. 114). The initial equal rates of breakdown in light and dark suggest that senescence is not triggered off by carbohydrate deficiency, although this factor may be important during the later stages of senescence in the dark.

In detached leaves, protein degradation leads to the accumulation of amino acids and amides in the leaf, since they cannot be exported, as is the case with attached leaves, although there may be accumulation at the base of the petiole where this is present. Thus, senescence can occur in detached leaves, or in leaf discs, even though there is a high level of amino acids within the tissues.

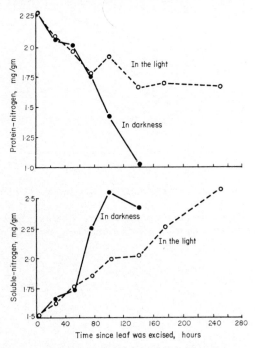

FIG. 114. Changes in protein and soluble nitrogen in *detached* tobacco leaves. Following excision, protein-N declines and soluble-N increases. Leaves kept in darkness show these changes more rapidly and are yellow and senescent after approximately 100 hours. (From H. B. Vickery *et al.*, *Connecticut Agric. Exp. Sta. (New Haven) Bull.* **374,** 557, 1935.)

Hence, under these conditions the reduced protein synthesis cannot be attributed to lack of amino acids. On the other hand, detached leaves show reduced capacity for protein synthesis, as indicated by the reduced capacity to incorporate ^{14}C-leucine into protein. Indeed, leaf discs which have remained in the dark for several days lose this capacity completely. Thus, there seems no doubt that the decline in protein content observed in detached leaves arises from a reduced capacity for protein synthesis.

It was observed that the rapid rate of protein breakdown in detached leaves is arrested if they are allowed to form roots (Fig. 115). Indeed, leaves which have been planted in soil and induced to form roots on the petiole will live for considerable periods. On these grounds, Chibnall suggested that roots must

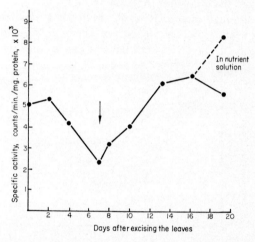

FIG. 115. Changes in capacity for protein synthesis (as measured by incorporation of ^{35}S-methionine into protein) in leaves of *Nicotiana rustica*. Note initial decline in capacity for protein synthesis, followed by recovery when roots appeared on petiole (indicated by arrow). (After von B. Parthier, *Flora, Jena*, **154,** 230, 1964.)

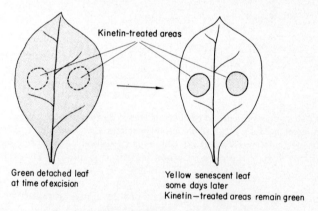

FIG. 116. Effects of kinetin on leaf senescence.

supply some "factor" which is necessary for the maintenance of protein synthesis in the leaves, but the nature of this "root factor" remained unknown. However, it was later found that kinetin will prevent the senescence of detached leaves if applied to them in aqueous solution. Thus, if a drop of kinetin solution is applied to a senescing tobacco leaf, then the area of leaf which received the kinetin will remain green, although the surrounding leaf tissue continues to yellow (Fig. 116). Similarly, if discs of radish or *Xanthium* leaves are placed on a solution of kinetin they still retain their green colour after the control discs (on water) have become fully senescent. Moreover, electron microscopic studies show that the kinetin-treated discs still have the normal structure of a green leaf.

Thus, senescence in detached leaves can be prevented either (1) by the presence of roots, or (2) by application of kinetin and other synthetic cytokinins. It was, therefore, natural to consider whether the natural "root factor" which delays senescence is an endogenous cytokinin. It is of great interest, therefore, to discover that kinins are present in the root exudate from sunflower, grape and other plants, and this suggests that leaves may depend on the supply of endogenous cytokinins from the roots, for the maintenance of a normal, green condition. We shall return to this question later (p. 269).

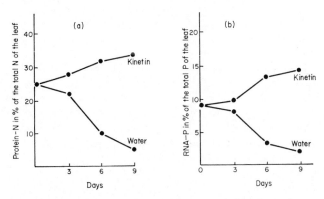

Fig. 117. Changes in (a) protein and (b) RNA content in kinetin- and water-treated halves of undivided leaves of *Nicotiana rustica*. (After R. Wollgiehn, *Flora, Jena*, **151**, 411, 1961.)

Not only does kinetin delay senescence of detached leaves, but it is found to cause a rapid increase in the rate of RNA and protein synthesis only a few hours after application (Fig. 117); incorporation of precursors, such as ^{14}C-adenine or ^{14}C-cytidine, into all fractions of RNA appears to be increased by kinetin. Since protein synthesis involves the RNA "machinery" of the cell, the effect of kinetin on protein synthesis is readily understood if its primary effect is upon RNA synthesis, but it may also delay senescence through its effect on the mobilization of metabolites (p. 267). The possible mode of action of cytokinins in stimulating RNA synthesis is discussed in a later chapter (p. 288).

With most species, gibberellins do not affect the rate of senescence of leaf discs, but in *Taraxacum, Rumex, Tropaelum* and a few other species, GA$_3$ is more effective in delaying senescence than kinetin. On the other hand, auxins are found to be relatively ineffective in delaying senescence in the leaves of herbaceous plants, but they are very active in this respect in the leaves of woody plants (p. 270), and in the pericarp tissues of beans (*Phaseolus vulgaris*).

Contrasting with the senescence-delaying effects of cytokinins and gibberellins, abscisic acid (ABA) *promotes* the senescence of isolated leaf discs of many species (p. 237). Moreover, ABA inhibits RNA and protein synthesis in leaf discs (Fig. 118). Its effects in senescence are therefore opposite to those of kinetin and GA$_3$.

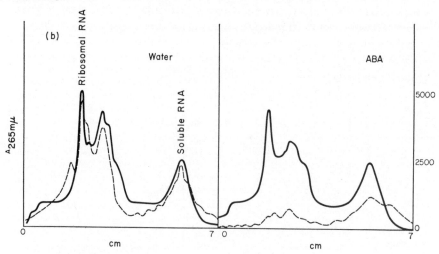

Fig. 118. Effect of abscisic acid on synthesis of RNA in leaf discs of radish (*Raphanus sativus*). Leaf discs were pre-treated with water or abscisic acid (10 mg/l) for 20 hours, and were then incubated with ^{14}C-cytidine (an RNA precursor) for 3 hours. After extraction the RNA was fractionated by electrophoresis on acrylamide gels. *Continuous line:* optical density at 265 nm. *Broken line:* radioactivity (in discharges per min). (L. Beevers, J. A. Pearson and P. F. Wareing, unpublished.)

Ethylene is another substance which may be important in senescence. It is now well established that ethylene plays an important role in the ripening of succulent fruits (p. 78), a process which is probably closely related to senescence in leaves and other organs. It may be of significance, therefore, that ethylene promotes the senescence of leaves, but how far it is involved in normal senescence processes is still not clear.

WHOLE PLANT SENESCENCE

As we have seen, in monocarpic species, such as wheat (*Triticum*), soybeans (*Glycine max*) and French beans (*Phaseolus vulgaris*), senescence of the whole plant is usually associated with the development of fruits. Thus, in the French bean plant, as the pods and seeds grow and approach their full size, there is perceptible yellowing of the leaves. When fully developed, the fruits yellow and ultimately lose water as they ripen, and at the same time the leaves and stems become progressively more senescent and in due course they die. Not only are plant senescence and fruit development closely correlated in time, but there does indeed appear to be a causal relationship between the two processes, as shown by the fact that removal of all the flowers and young fruits from bean plants greatly delays senescence of the remainder of the plant. Moreover, the same effect is observed if, instead of removing the whole fruits, only the seeds are removed surgically from the pods. Thus, the presence of developing seeds appears to control plant senescence, suggesting that some "signal" is sent out from the developing seeds, which regulates senescence in other parts of the plant.

Now, the developing seeds of bean increase steadily in dry weight, and food reserves, including protein and starch, accumulate in the developing cotyledons. It is known that, at the same time, the breakdown of proteins and carbohydrates is occurring in the leaves and that amino acids, sugars and other metabolites are transported to the developing fruits. Similar effects are seen clearly in other species. It was, therefore, suggested by Molisch that developing seeds cause senescence in other parts of the plant by mobilizing and accumulating not only carbohydrates but also amino acids and other substances derived from the leaves.

The accumulation of reserve materials into fruits, tubers and other storage organs is well known. Similarly, actively growing meristematic regions, such as young leaves, are known to be able to mobilize nutrients from other parts of the plant. Such centres of mobilization are often referred to as "sinks" for nutrients. The manner in which mobilization by such sinks is achieved is not understood, since the mechanism of phloem transport is itself still not understood. However, in recent years increasing evidence has appeared indicating that growth hormones may play an important role in mobilization effects. Thus, when kinetin is applied to a small area on a senescing tobacco leaf, it is found that radioactive amino acids applied to the other parts of the leaf are accumulated at the point of application of kinetin. Thus, metabolites appear to be "attracted" towards regions which have been treated with kinetin. It is possible that this effect of kinetin on the mobilization of metabolites is a secondary one, arising from the stimulation of protein synthesis at the point of kinetin application, so that a metabolic "sink" is created, which in turn causes movement of metabolites towards this point. However, Möthes was able to show that if amino-isobutyric acid (which is not a naturally occurring substance and is not incorporated into proteins) is applied to a part of the leaf away from the point of kinetin application, it accumulates in the latter region, just as the natural amino acids do. Thus, on this evidence it would seem that kinetin may affect the movement of metabolites independently of any effect it has on protein synthesis.

There is evidence that auxin-directed transport may also play an important role in the movement of metabolites in ripening bean plants. We have seen (p. 125) that metabolites tend to accumulate in regions of high auxin content and it is known that developing bean seeds are rich sources of auxin. Thus, it is possible that the movement of reserve materials from the leaves into the developing seeds may be an example of auxin-directed transport. Evidence in support of this hypothesis has been obtained in the following experiment. The fruits were removed from young bean plants, and a decapitated peduncle (fruit stalk) was allowed to remain on each plant. In some of the plants indole-acetic acid (IAA) in lanolin was applied to the peduncle stump and in other plants (controls) plain lanolin was applied. It was shown that when radioactive phosphorus, ^{32}P (in the form of orthophosphate) was applied to the stem bases, it very rapidly moved up into the peduncles to which IAA had been applied, but very little moved into the peduncles to which only plain lanolin was applied.

The hypothesis that plant senescence is brought about by the mobilization of nutrients by developing seeds would seem to be consistent with many observations, and if we postulate that the leaves become depleted of amino acids as a result of their export to the seeds, we have essentially the same hypothesis as was suggested above for the sequential senescence of leaves. However, there are certain observations which are difficult to reconcile with this hypothesis. Thus, it is found that in spinach (*Spinacia oleracea*), which is a dioecious species (with separate male and female plants), senescence of the *male* plants follows flowering, just as in female plants, although such male plants do not, of course, carry any developing fruits; moreover, removal of the male flowers delays senescence of the leaves. Moreover, in an experiment with *Xanthium pensylvanicum*, it was found that if all buds were removed from plants before exposure to short days, so that no flowers could be formed, the leaves of these plants later senesced at the same time as did those of plants which had been allowed to form flowers and fruit. On these and other grounds, the hypothesis that plant senescence can be explained simply in terms of the mobilization of nutrients by developing seeds is rejected by many workers.

Since protein synthesis is controlled by the RNA apparatus of the cell, it is entirely possible that the presence of developing seeds brings about protein breakdown in the leaves through an effect upon their RNA metabolism. For example, some fractions of RNA undergo rapid turnover, and it may be that certain RNA precursors are exported from the leaf to the developing seeds, and so are not available for reincorporation into RNA in the leaf. A different explanation has been suggested, as follows. We have seen that leaves undergo rapid protein and RNA breakdown when they are separated from the plant and it has been suggested that they are dependent on a continuous supply of cytokinin from the roots for normal RNA metabolism. Now, seeds are found to be rich in cytokinins. It is not known whether these cytokinins in seeds are synthesized there, or whether they are mobilized there from other parts of the plant. If the latter were the case, then it is possible that cytokinins produced in the roots are directed away from the leaves in the presence of developing seeds, with the result that the normal maintenance of the RNA apparatus of the leaf is not possible and hence protein synthesis is prevented, as in a detached leaf. One observation which is against this latter hypothesis is the fact that the senescence of attached leaves cannot normally be arrested by application of kinetin, suggesting that they are not deficient in cytokinin and hence their senescence must be due to some other cause. Thus, it would seem possible that the causes of senescence in a detached leaf are not the same as those of an attached leaf which undergoes sequential senescence or in response to fruiting.

It will be clear that although a number of interesting approaches to the problem of senescence in plants are being made, it is too early to be able to present a single overall hypothesis which will account for all the facts.

SYNCHRONOUS LEAF SENESCENCE

The synchronous senescence of leaves seen in deciduous woody plants in the autumn or "fall" is so striking that it has given its name to this season of the year, in America at least. This type of leaf senescence differs rather markedly from sequential senescence in two respects. Firstly, it is primarily controlled by environmental rather than "internal" factors, such as competition between young and old leaves. Secondly, it appears to involve rather different hormonal factors.

Two environmental factors appear to be involved in determining the onset of senescence in deciduous woody plants, namely, daylength and temperature. It has long been known that short days tends to promote leaf senescence in woody plants such as *Liridodendron tulipifera* and *Ailanthus altissima*. Moreover there have frequently been reports of delayed leaf fall in trees growing near street lights, so that they were exposed to long days as the natural daylength shortened. However, if seedlings of woody plants are rendered dormant by placing them under short-day conditions in a warm greenhouse, many species are found to retain their leaves in a green, healthy condition for several weeks at least. Thus, it seems likely that normal leaf fall is determined by short days in association with the low temperatures occurring in the fall, but precise experimental data on this matter are still lacking.

Although environmental factors are very important in controlling synchronous leaf senescence, influences from the rest of the plant are evidently also operating, since if a disc is nearly cut out of a cherry leaf by means of a cork borer, leaving only a small connection with the rest of the lamina, this disc remains green long after the remainder of the leaf has become senescent. Thus, it would seem that synchronous, as well as sequential senescence, depends upon the export of materials from the leaf.

We have seen that the senescence of excised leaves or leaf discs of herbaceous plants can be delayed by kinetin, and sometimes by gibberellin, but not by auxins. In woody plants, however, the reverse appears to be true. Thus, if a drop of the auxin, 2,4-D, is applied to cherry leaves in the fall, the areas receiving the auxin remain green long after the remainder of the leaf has become yellow. Similar results may be obtained with detached leaves of woody plants at all times of the year. Gibberellin will delay senescence of leaves of ash (*Fraxinus excelsior*). Just as application of kinetin and gibberellin enables the leaves of herbaceous plants to maintain their capacity for RNA and protein synthesis, so does 2,4-D for leaves of woody plants. Thus, it would seem that endogenous auxins probably play an important role in the natural senescence of leaves of trees. It is very likely that the influence of daylength on leaf fall is exerted through its effect on endogenous hormone levels, since there are lower levels of endogenous auxins and gibberellins, and higher levels of inhibitors in the leaves of woody plants under short days than under long days (Fig 103).

We know little about factors controlling leaf senescence in evergreen broadleaved trees and conifers. Some evergreen species may retain their leaves for several years. In some cases senescence of the older leaves occurs when a new suite of leaves is put out, suggesting that possibly competition between old and new leaves is important in such species.

FURTHER READING

General

Leopold, A. C. *Plant Growth and Development*, McGraw Hill, New York, 1964.
Leopold, A. C. Senescence in plant development, *Science*, **134**, 1727, 1961.

More Advanced Reading

Articles by H. W. Woolhouse, E. W. Simon, R. Wollgiehn, D. J. Osborne, P. F. Wareing and A. K. Seth, and D. J. Carr and J. S. Pate, in Aspects of the Biology of Ageing, *Symp. Soc. Exp. Biol.*, **21**, 1967.

CHAPTER 13

The Control of Development

WE HAVE now considered plant development from several rather different standpoints. In the earlier chapters we considered what can be found out about morphogenesis and differentiation from the approach of the experimental anatomist and morphologist. Then we examined the action of growth hormones and discussed their roles in various aspects of growth and development. Finally we considered the physiology of flowering, senescence and dormancy, bringing in the effects of external factors, such as daylength, and internal factors, especially hormones.

In the present chapter we shall examine some more general aspects of development, particularly from the standpoint of what might be called "molecular genetics". So far we have paid little attention to the genetical aspects of development, and yet every organism is, by definition, the product of the interaction between its genetic potentialities and the environment, and in the final analysis development has to be described in terms of the activities of genes.

GENES AND DEVELOPMENT

The fact that developmental processes are basically gene-controlled is self-evident, since genetic variations are known which affect almost every aspect of development, ranging from external morphology (leaf and fruit shape, flower colour, etc.) and internal anatomy, to physiological characters, such as growth rate, flowering time and length of dormancy period. Thus, there seems no doubt that the pattern of development followed by any individual is determined primarily by the "programme" laid down in its genetic code. During development there is progressive differentiation of organs and tissues, giving rise to a wide range of different types of cell. Not all the genes of the total gene complement are operative all the time and in all parts of the plant, however. Thus, the genes controlling flower development are apparently not normally operative in the embryo, or during leaf development. However, we know that the cells of the leaf contain the genes controlling flower development, since a new plant may be regenerated from a leaf in certain species, and this plant is found to possess a complete set of genes, i.e. the differentiated cells of the leaf still show *totipotency*. Thus, in plants, differentiation apparently does not involve any genetic, i.e. inheritable, differences between the nuclei of the various types of cell and tissue. We must conclude, therefore, that there is some means of switching genes "on" and "off" at different stages of development, and since development is a very

Fig. 119. The alga, *Acetabularia*, consists of a single giant cell, the nucleus of which is located in the rhizoid region.

orderly process there must be some mechanism for determining the sequence of the gene-switching processes. Therefore, we shall have to discuss in some detail how gene activation is achieved, and what controls the sequence of gene-switching.

However, the genes do not operate in isolation, and the cytoplasm also plays a vital role in differentiation, and hence in controlling gene activation. This is seen, for example, in the role of unequal division in differentiation (p. 54), where cytoplasmic differences between two ends of a parent cell appear to determine the patterns of differentiation in the two daughter cells. Moreover, the visible manifestations of differentiation between cells occur primarily in the cytoplasm and cell wall. Thus, the observable differences between, say, a root hair and a mesophyll cell relate primarily to the shape and character of the wall, and the relative development of cytoplasmic structures such as plastids. It is clear that differentiation involves an interplay between nucleus and cytoplasm, and that the cytoplasm both affects gene activity and is itself the site of the results of such activity. These conclusions are well illustrated in the unicellular alga *Acetabularia*.

Acetabularia consists of a single, giant cell, which has a rhizoid and a cylindrical stalk (Fig. 119). The stalk grows in length at its apical end and ultimately forms a circular cap consisting of a considerable number of sections called "rays", in which large numbers of cysts are ultimately formed and released; the cysts give rise to isogametes after shedding. There is a large central vacuole, with a boundary layer of cytoplasm inside the wall, and the single nucleus lies in the cytoplasm at the rhizoid end.

Acetabularia shows considerable capacity for regeneration and if the stalk is cut off above the rhizoid the latter will regenerate a new stalk and cap. However, even parts lacking a nucleus can regenerate. For example, if a young stalk which has not yet developed a cap is separated from the rhizoid, so that it lacks a nucleus, it will survive for a long time and will ultimately form a cap.

Different parts of the cell may be grafted together. If the apical part of the young stalk is cut off and pushed over the basal part of another plant containing a nucleus, it will continue to grow and develop normally. Grafting may also be carried out between various species of *Acetabularia*, which differ in the shape of the cap. If an anucleate section of one species is grafted to a nucleate base of another species, the regenerated plants show mainly the characters of the species which supplied the nucleus.

These various results seem to show that development is ultimately under the control of the nucleus, but that there are long-lasting effects of the nucleus on the cytoplasm, which can continue and complete morphogenesis even if the nucleus is removed. It has been suggested that certain "morphogenetic substances", possibly messenger-RNA (p. 274) formed in the nucleus, are present in the cytoplasm and persist over long periods.

HOW GENES ACT

The spectacular recent advances in our understanding of the way in which genes act, and which are frequently referred to as "molecular biology", are now widely known and hence need only be summarized briefly here.

The first important clue, arising from genetic studies with bacteria and fungi such as *Neurospora*, was that there is a relationship between enzymes and genes; that is to say, the synthesis of each enzyme found in the cytoplasm is dependent upon and controlled by specific genes in the nucleus.

How does a gene (i.e. a specific segment of DNA) determine the sequence of amino acids in a given enzyme protein? We now know that each amino acid is coded for in the sequence of base pairs (adenine–thymine, guanine–cytosine) which are present in the double helix of the DNA molecule, and that a group of three successive base pairs (triplet or codon) acts as a code for one amino acid. The next three nucelotides code for the next amino acid, and so on.

The information coded in the DNA thread is apparently "transcribed" to a single-stranded ribose nucleic acid (RNA) thread, referred to as "messenger"-RNA (m-RNA). It is thought that in the messenger-RNA the sequence of bases

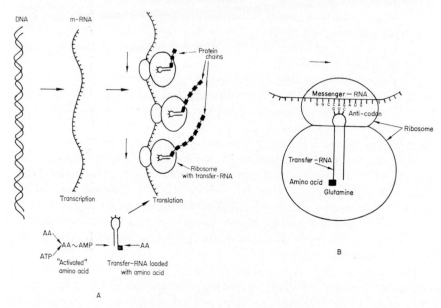

Fig. 120. Scheme to illustrate the main stages in protein synthesis (see text.)

is complementary to that in the DNA thread, so that if guanine is present in the DNA, cytosine will be laid down in the m-RNA thread and vice versa. On the other hand, thymine is not present in RNA, so that where adenine occurs in the DNA thread, uracil (not thiamine) apparently occurs in the m-RNA thread. In this way, the information coded in the DNA can be imprinted in the m-RNA in a complementary manner which resembles a die and a mould. It is generally assumed that the m-RNA leaves the nucleus and passes out into the cytoplasm where *ribosomes*, particles containing a different form of RNA, become attached to it, but there is little direct evidence for this assumption.

The m-RNA evidently controls the sequence in which amino acids are assembled in the process of protein synthesis (Fig. 120). The amino acids first become "activated" by reacting with adenosine triphosphate to form a complex between the amino acid and adenosine monophosphate (AMP) through a "high energy" bond between the carboxyl group of the amino acid and the phosphate of AMP. This process is catalyzed by an "activating enzyme", which is specific for a given amino acid. The activated amino acid (AA ~ AMP) remains attached to the enzyme, and then becomes linked to a third form of RNA molecule, known as *transfer-RNA* (t-RNA). Each amino acid is attached to its own specific t-RNA. The combined amino acid and t-RNA become associated with the m-RNA; starting from one end, successive groups of base triplets in the m-RNA become combined with t-RNA molecules (to each of which an amino acid is attached). A specific region of the t-RNA is complementary to a particular triplet ("codon") on the m-RNA, and hence is referred to as *the anticodon*.

It is generally assumed that the messenger-RNA is held by the ribosome so that two successive triplets can be recognized simultaneously by the appropriate t-RNA molecules. The two amino acids are then attached to each other by the formation of a peptide bond, in a manner which is not yet fully understood. By a repetition of this process a chain of various amino acids is formed. The original DNA thread passes on its code to the m-RNA, which in turn is recognized by the complementary sequence in the various t-RNAs. In this way the sequence of bases in the DNA determines the sequence of amino acids, which, in turn, determines the structure of the enzyme protein. The information originally coded in the DNA becomes "transcribed" in the formation of m-RNA, and is "translated" in the process of protein synthesis.

Our knowledge of protein synthesis is based mainly upon studies on bacteria and animal cells, but there is little doubt that the process is essentially similar in all organisms. The occurrence of transfer and ribosomal RNA in higher plants is well-established, but it has been more difficult to demonstrate the presence of m-RNA. Indeed there is considerable uncertainty regarding the molecular size and composition of m-RNA, which has not yet been isolated in a pure form and unequivocally characterized for any higher organism.

CONTROL OF GENE ACTIVITY IN BACTERIA

Having reviewed briefly how genes control enzyme synthesis, we now have to consider how genes are switched on and off. Bacteria have proved very favourable material for the study of this problem.

Some enzymes of bacteria are present all the time and are said to be *constitutive*. On the other hand, other enzymes are only formed when their substrate is present in the external medium. For example, when the bacterium, *Escherischia coli*, is grown in the absence of a galactoside (i.e. a compound containing the sugar, galactose, linked to another, non-sugar, molecule), only traces of the enzyme β-galactosidase are formed, but as soon as a galactoside is added, the rate of synthesis of this enzyme increases enormously. This type of enzyme is said to be *inducible*. Removal of the substrate results in the almost immediate cessation of enzyme synthesis.

By contrast with this process of enzyme induction is the *repression* of enzyme synthesis seen in other enzyme systems. For example, when *E. coli* is grown in the absence of the amino acid histidine, the enzymes involved in the production of histidine are actively synthesized. As soon as histidine is added to the medium the enzymes cease to be synthesized. In this instance, therefore, there is *repression* of enzyme synthesis. This phenomenon is called "end-product repression", since the product of the sequence of reactions, in this case histidine, inhibits the formation of the enzymes concerned with its own biosynthesis. The result is that if the end-product is supplied from outside, it inhibits its own synthesis, and hence the cell ceases to make any more of the compound.

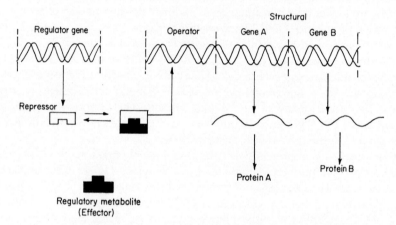

Fig. 121. Scheme illustrating the control of protein synthesis according to the theory of Jacob and Monod (see text).

The phenomenon of enzyme induction and repression led Jacob and Monod to propose a general theory for the control of gene activity in bacteria, which we shall now describe briefly. We know that the structure of a specific enzyme is determined by a specific gene or segment of DNA, and we have seen that the information coded in the DNA is transferred to the cytoplasm via messenger-RNA, and is "translated" during protein synthesis. The gene which determines the structure of a specific enzyme protein is referred to as a *structural gene*. The synthesis of m-RNA is assumed to be initiated only at certain regions or *operators* of the DNA strands (Fig. 121). In some instances a single operator may control the transcription of several adjacent structural genes into m-RNA. The segment of DNA thus controlled by a single operator is referred to as an "operon", which may contain one or several structural genes.

The rate of transcription of structural genes is controlled by other genes referred to as *regulator genes*. A regulator gene forms a cytoplasmic product or *repressor*. The repressor formed by a given regulator gene is assumed to have an affinity for certain specific operator genes, with which it binds. This combination blocks the production of m-RNA by the whole operon controlled by the operator, and therefore prevents the synthesis of the specific proteins controlled by the structural genes of the operon.

A repressor (R) has the property of reacting with certain small molecules, called "effectors" (F) expressed as follows:

$$R + F \rightleftharpoons RF$$

In inducible systems only the R form of the repressor is active and blocks the transcription of the operon. The presence of an effector (or inducer) inactivates the repressor and therefore allows messenger-RNA synthesis to proceed. By contrast, in repressible systems only the combined RF form of the repressor is active. Synthesis of m-RNA by the operon, allowed in the absence of the effector (or repressing metabolite), is, therefore, prevented in its presence. These ideas are summarized in Fig. 121.

The repressor may be an "allosteric" protein, with two active sites, one able to react with the operator and one to react with the inducing or repressing molecule. It is supposed that when the repressor combines with an inducer its shape is deformed somewhat, so that it can no longer react with the operator, which is thereby "derepressed".

There is much evidence in support of the existence in bacteria of the type of regulator system postulated by Jacob and Monod, but there is no evidence to suggest that it exists also in higher plants. Moreover, even in bacteria the Jacob and Monod hypothesis apparently applies to the regulation of only a few enzymes. However, genes are known in maize which appear to show some of the characters of regulator genes. Thus, a locus in maize known as "Activator" (Ac), appears to be a master locus for a second locus, "Dissociation" (Ds), which is unable to function in the absence of Ac. Ds, in turn, affects the expression of a number of other genes, and hence is analogous to an "operator" gene in bacteria, while Ac may be regarded as the "regulator".

For example, under certain conditions Ds causes C, which gives coloured aleurone in the grains, to behave as if it were the colourless recessive allele, c. Thus Ds appears to "repress" the action of C. On the basis of these and other observations, McClintock has postulated that the chromosomes of higher plants contain both genes and "controllers", which regulate the action of the genes.

It is not difficult to construct models, based upon the Jacob and Monod theory, which would explain several features of differentiation in plants, as we shall see later (p. 284).

CONTROL OF GENE ACTIVITY IN HIGHER PLANTS

We do not yet know whether the bacterial type of control system is operative in higher organisms. On the other hand, evidence is accumulating which suggests that a somewhat different mechanism may also be involved in the control of gene activity in higher plants and animals.

It has long been known that the nuclei of plants and animals contain not only DNA, but also significant amounts of protein, most of which is histone, a type of protein which is basic in nature, due to the high content of the basic amino acids, lysine and arginine. The possible role of histones in repressing DNA has been studied, particularly by J. Bonner and his co-workers, by using preparations of chromosomal material referred to as "chromatin". Chromatin is prepared by disruption of the tissue by homogenization, filtering and selective sedimentation by centrifugation. In this way relatively pure interphase chromosomal material may be prepared and its properties studied *in vitro*. Such chromatin contains DNA, histone, a small amount of non-histone protein and a small amount of RNA. It may also contain the enzyme RNA-polymerase, which is responsible for the linking of nucleotides during the synthesis of m-RNA.

Isolated chromatin is capable of synthesizing RNA when supplied with the nucleoside triphosphates of the four RNA bases, guanine, adenine, cytosine and uracil. It is thus possible to study the RNA-synthesizing capacity of DNA from various types of plant material. For example, in the pea plant a specific protein, a globulin, is made and stored in the cotyledons as reserve material and is not made in other parts of the plant. If chromatin is prepared from cotyledons and supplied with bacterial ribosomes and various other substances, then it is able to make a messenger-RNA, which in turn leads to the synthesis of globulin. The chromatin from pea buds, on the other hand, makes no globulin under these conditions. Thus the gene which codes for globulin is active in chromatin from cotyledons, but not in the chromatin from other parts of the plant. This finding gives direct evidence for the selective repression of the DNA from different tissues.

It has been shown that chromatin extracted from dormant potato buds shows lower RNA-synthesizing ability than chromatin from non-dormant buds, suggesting that possibly dormancy involves rather extensive repression of DNA. Similar results have been obtained with chromatin prepared from dormant and non-dormant hazel seeds.

These observations do not, of course, show that histone is involved in the repression mechanism, the chief evidence for which is provided by the finding that when the histone is removed from chromatin, the deproteinized DNA is found to have very much greater RNA-synthesizing activity than chromatin not so treated. Moreover, whereas native chromatin from pea-buds cannot make globulin, deproteinized pea-bud chromatin can do so. Evidently removal of the protein, including the histone, from pea-bud chromatin derepresses the genes for globulin synthesis.

One difficulty for the hypothesis that histones act as gene-repressors is the fact that relatively few types of histone are known to occur naturally, whereas there is a large number of genes, each of which would seem to require a specific repressor. This difficulty remains unresolved at present, although it has been suggested that the small amount of RNA present in chromosomes may be bound to the histones and confer specificity upon them. A further problem is the fact that histones are not present in the bacterial chromosome and yet there is effective gene activation and repression in bacteria. Indeed, the repressors in bacteria appear to be specific, non-histone proteins.

PROGRAMME SELECTION IN DEVELOPMENT

So far, in this chapter, we have been considering some of the possible mechanisms which control the activation and represssion of genes. We now have to consider these processes in relation to development of the organism as a whole.

We have seen that development of a flowering plant involves a series of successive stages, starting with the differentiation of the embryo into root and shoot, followed by the formation of organ initials, and ending with tissue differentiation within the individual organs. Superimposed on this pattern are the major changes in development represented by the transition to the flowering phase and the onset of dormancy. The orderly manner in which the successive stages follow each other in a regular sequence is one of the most striking features of development. This phenomenon is well illustrated in the development of a flower, where not only is there a regular sequence in the initiation of the various flower parts (perianth, stamens and carpels), but within each of these there is an equally regular pattern of development.

The transition from one stage of development to the next would seem to involve a process of successive gene activation, in which certain previously repressed genes come into action and others become or remain repressed. Direct evidence of successive gene activation is provided by insect development. As is well known, the cells in some organs of *Drosophila* and certain other flies have "giant" chromosomes, formed by the repeated replication of the DNA strands, without nuclear division. These giant chromosomes thus each consist of large numbers of single chromosome threads lying parallel to each other and aligned so that corresponding regions of the various threads are opposite to each other, to give a characteristic banded appearance. Each band appears to correspond to a single gene or operon. At certain stages of development of the insect one or more of these bands swells up and forms a "puff" of what is apparently RNA. It would seem that the occurrence of puffing indicates that a particular gene is active at that time. Different tissues have characteristic puffing patterns, and these occur at specific periods of development. Moreover, if the moulting hormone, ecdysone (p. 85), is administered to a larva the puffing pattern changes rapidly, as the insect enters a new phase of development. We have no direct evidence for successive gene activation during development in plants, but it seems very likely, on *a priori* grounds, that this does occur.

If development involves an orderly sequence of activation and repression of large numbers of genes, this raises the question as to how such gene control is achieved. Clearly specific genes must be activated in the right cells at the right time.

The successive stages of development can be regarded as a process of "programme selection", in which there is divergence into alternative pathways of further development at various critical points of time and space. This divergence may occur at the cellular level, as when two daughter cells of a common mother cell show divergent patterns of differentiation (p. 54), or it may occur in the differentiation of organs, or even in the shoot apex as a whole, as in the transition from the vegetative to the flowering phase. Furthermore, we saw that once an organ, such as a leaf primordium, passes a certain stage of development, it becomes irreversibly "determined" as a leaf (as opposed to a bud) and cannot normally then be converted into any other structure (p. 37).

As stated earlier (p. 23), Waddington has described similar phenomena in animal embryo development as "canalization of development" and has pointed out that once a given organ has become canalized into a particular pathway of development it is difficult to divert it into another pathway. He suggests that the situation can be visualized by imagining a series of alternative valleys or grooves, which he describes as forming the "epigenetic landscape" (Fig. 122). If a ball is placed at the top of the slope and allowed to run down, it will be diverted into one of the alternative pathways at various stages, but once it has entered a given valley it cannot easily be displaced over the "hill" into a neighbouring valley. Similarly, by analogy we may say that once a plant structure has entered a particular developmental pathway it becomes "determined" and it is then very difficult to change the pattern of subsequent events, except to modify them within certain restricted limits.

The determination of an organ or a tissue would seem to imply that the canalizing processes involve changes which are not easily reversed. Moreover, the "determined" state can apparently be transmitted even through successive cell divisions, since in leaf development many cell divisions occur after the time at which the primordium becomes determined. The phenomena of canalization and determination thus pose a number of questions: (1) How is entry into a particular pathway decided? (2) How is the regular sequence of activation of genes achieved? (3) What does determination involve, in terms of gene activation and repression?

At the present time we have no decisive answers to any of these questions and we can only speculate as to the types of mechanism which might be involved.

The selection of alternative pathways of development is well illustrated in the results of unequal cell division, where two daughter cells of the same parent cell follow divergent patterns of differentiation (p. 54). An example is seen in the initial division of the zygote, both in lower and in higher plants, as a result of which differentiation of the plant body into shoot and root (or thallus and rhizoid) regions is determined. We have seen that in a cell about to undergo unequal division it is frequently possible to recognize differences in the density

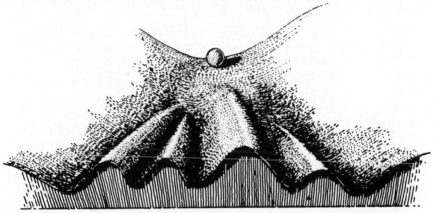

Fig. 122. The "epigenetic landscape". The various parts of a developing organism have in front of them a number of possible pathways of development, and any particular part will be switched into one or other of these potential paths. (From C. H. Waddington, *Principles of Development and Differentiation*, MacMillan, New York, 1966.)

of the cytoplasm between the two ends of the cell, before division occurs. This fact suggests that there is asymmetric distribution of certain substances in the cytoplasm which become differentially distributed between the two daughter cells, and determine the subsequent differences in gene activity in these cells. While such unequal divisions may be important in the differentiation of organs and tissues in mosses and other lower plants, such as *Selaginella* and certain species of *Equisetum* (p. 56), this mechanism appears to be of only limited importance in the higher plants, where it may be involved in the early differentiation of the embryo into root and shoot poles, and in the later stages of differentiation, as in the formation of stomatal guard cells (p. 54). But in the shoot apex of the flowering plant, unequal divisions do not appear to be the basis of differentiation into leaves, buds and the various tissue layers of the stem.

The development of the embryo appears to involve canalization into several major pathways, as a result of which there is determination of root and shoot regions, and with the establishment of an organized shoot apex the basis is laid for the initiation of various organs (leaf, bud, stem), which is maintained throughout the subsequent vegetative phase of the plant. It is possible that the establishment of the major pathways involves the activation of a series of "master genes" controlling the activities of a large number of subordinate genes, which become activated during the subsequent differentiation of an organ.

The regular sequence of changes seen in the development of an organ such as a leaf or a flower strongly suggests that once these particular pathways have been entered, then all the subsequent stages follow inexorably, as a sort of "chain reaction" in which the attainment of one stage seems to trigger off the next one. If this latter type of mechanism is involved, then it would imply that the pattern of developmental events is predetermined, at least for certain organs, and that development is primarily controlled by an internal mechanism, as if the organism is an internally programmed piece of equipment which, once it is switched on, will automatically go through a regular sequence of activities, each of which is triggered off by the preceding one.

A mechanism of this type has been suggested for flower development. It is postulated that once the transition to the flowering stage has begun, a gene complex A is activated in the first formed primordia, and these genes produce an inducer X which moves to the next set of primordia and there activates a gene complex B, and so on through the different classes of organ. This hypothesis postulates the occurrence of short-range inter-cellular "messengers" or "hormones", but the existence of such substances has yet to be demonstrated.

By contrast with the "internally programmed" type of development we have just been considering, we have already seen that certain stages of the life cycle, notably flowering and bud dormancy, may be controlled by environmental factors such as daylength and temperature. In plants showing photoperiodic control of flowering, it is clear that the onset of flowering is not rigidly predetermined as part of a sequence of internally programmed changes, but may be controlled by a factor of the external environment. In this case the daylength conditions may control gene switching through phytochrome.

Thus, we apparently have two rather contrasting types of control mechanism. Internal programming, such as we see in leaf and flower development, appears to be particularly characteristic of organs of determinate growth, whereas environmental control appears to be more common with changes involving the shoot apex, as in the initiation of flowers and dormant resting buds. However, this distinction between the development of organs of determinate and indeterminate growth patterns is not absolute, since light also plays an important part in normal leaf development (p. 180).

BIOCHEMICAL DIFFERENTIATION IN HIGHER PLANTS

Ultimately the processes occurring during cell differentiation are terminated and the cell reaches the "steady state" of the mature condition, in which metabolism is maintained continuously (except, of course, in the case of non-living cells such as those of the xylem). The visible manifestations of differentiation include variations in the development of the cell wall and of certain cytoplasmic organelles, such as plastids. It is clear that differentiation must also extend to certain aspects of metabolism when it is remembered that some tissues are specially adapted for particular functions, such as photosynthesis, secretion and storage of reserve materials. Such differentiation almost certainly involves differences in enzyme production, which in turn implies the maintenance of differences in gene activation and repression between various cells even in the mature state.

Many basic metabolic pathways are probably operative in all living cells of the plant. This must apply to the main pathways involved in the respiratory breakdown of carbohydrates, for example. On the other hand, there is much evidence that various tissues differ in their biosynthetic abilities. For example, it is found that the isolated roots of many species require the supply of certain vitamins, including thiamin, pyridoxine and nicotinic acid when grown in sterile culture. It appears that in the intact plant these vitamins are synthesized in the shoot and are supplied to the roots. Similarly, callus cultures of certain plant tissues, including tobacco pith tissue, require to be supplied with auxin and cytokinin in order to maintain cell division in culture (p. 152). Evidently tobacco pith cells do not have the ability to synthesize auxin and cytokinin and hence require an exogenous supply. However, this inability of pith cells to synthesize the two hormones is not due to any permanent loss of the potentiality to do so, as is shown by studies on the bacterial disease known as "crown gall" (p. 152). The bacterium apparently produces a substance, known as "tumour inducing principle", which brings about the transformation of normal plant tissue into tumour tissue, and once transformed it will continue to grow as tumour tissue indefinitely. As a result of infection the normal tissue has apparently undergone a profound and stable change. It is found that the tumour tissue will grow actively in sterile culture without the addition of auxin and cytokinin, indicating that it can now synthesize its own hormones. Thus, the capacity for synthesis is not irreversibly lost in the normal tobacco pith cells, but evidently the genes controlling the formation of the enzyme involved in hormone biosynthesis are repressed, but they become derepressed in some way by the tumour inducing principle. (A different interpretation is suggested, however, by the recent report that transformation of the plant cells involves the incorporation of part of the bacterial genetic "information" into the plant cell, so that the latter can now synthesize certain guanidines (purines).)

The inability of roots to synthesize certain vitamins, and of tobacco stem tissue to synthesize auxins and cytokinins, provides rather strong evidence that cell differentiation involves the activation of certain genes and the repression of others. It would be of interest to know whether the meristematic cells of the shoot apex of tobacco have the ability to synthesize cytokinins; if so, then it would seem that one of the processes occurring during cell differentiation in the stem is the repression of the enzymes responsible for auxin and cytokinin synthesis. Such a change in synthetic ability might, indeed, explain the transition from cell division to cell expansion seen in the apical region of both shoot and root. On the other hand, all cells of the shoot may lack the capacity for biosynthesis of cytokinin and may depend upon supply of this hormone from the roots (p. 265).

We do not know how these permanent differences in biosynthetic ability are maintained, but Jacob and Monod have shown that it is not difficult to construct

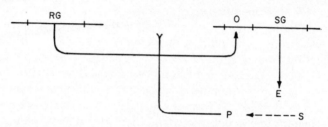

Fig. 123. Hypothetical model of a circuit for control of enzyme synthesis in which the inducing substance is the product of the controlled enzyme. Synthesis of enzyme E, genetically determined by the structural gene SG, is block edby the repressor synthesized by the regulator gene R.G. The product P of the reaction catalysed by enzyme E acts as an inducer of the system by inactivating the repressor. O, operator; S, substrate. (From F. Jacob and J. Monod, *Cytodifferentiation and Molecular Synthesis* (Ed. M. Locke), Academic Press, New York and London, 1963.)

"model" circuits, based upon the bacterial repression concepts, which would result in certain genes being permanently switched on. One very simple circuit is shown in Fig. 123, in which the inducer is not the substrate but the product of the controlled enzyme system. Such a system is known to occur in bacteria. In the absence of an exogenous inducer, the enzyme will not be produced, unless already present, but temporary contact with an inducer will cause the system to become "locked" indefinitely, at least so long as either substrate or product is present. Suppose, for example, that the operon in Fig. 123 is responsible for the biosynthesis of a hormone, such as cytokinin or gibberellin, and that in a dormant seed the operon is repressed. Then a single treatment with exogenous hormone will activate the operon and from then on the seedling would be able to synthesize its own hormone. The stimulation of endogenous auxin synthesis by a single application of exogenous IAA in fruit set (p. 117) may provide a similar example. It is quite easy to construct other models which would account for the permanent suppression of certain enzymes. Thus, in the model illustrated in Fig. 124, the product of one enzyme acts as an inducer for the other, so that the two enzymes are mutually dependent. One could not be synthesized in the absence of the other, and inhibition of one enzyme or elimination of its substrate, even temporarily, would result in the permanent suppression of both.

There is no evidence that determination and differentiation involve this type of mechanism, but these circuits do illustrate that it is possible to construct models which would account for the observed facts, within the concepts of bacterial repressor mechanisms.

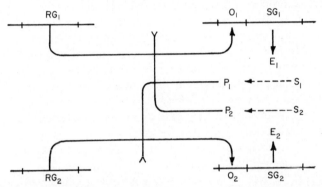

FIG. 124. Model circuit in which the product of one enzyme acts as an inducer of the other. Synthesis of enzyme E_1, genetically determined by the structural gene SG_1, is blocked by the repressor synthesized by the regulator gene RG_1. Synthesis of another enzyme E_2, controlled by structural gene SG_2, is blocked by another repressor, synthesized by regulator gene RG_2. The product P_1 of the reaction catalysed by enzyme E_1 acts as an inducer for the synthesis of enzyme E_2, and the product P_2 of the reaction catalysed by enzyme E_2 acts as an inducer for the synthesis of enzyme E_1. O, operator; S_1, S_2, substrates. (From F. Jacob and J. Monod, *Cytodifferentiation and Molecular Synthesis* (Ed. M. Locke), Academic Press, New York and London, 1963.)

Not only are differences in biosynthetic ability observed in mature, non-dividing cells, but they can also be maintained through cell division in certain tissues. For example, callus tissue from juvenile and adult parts of the ivy vine show differences in growth rate and in the capacity to form roots even when maintained under identical conditions in tissue culture. A number of other examples are known in which callus cultures derived from different tissues of the parent plant continue to show morphological and metabolic differences for considerable periods after they have been brought into culture.

There are also a number of clear instances of the maintenance of a differentiated state through many cell divisions in normal development. For example, we have seen that in winter rye the vernalized condition can be transmitted apparently without dilution through many cell divisions. The differentiation into independent, free-living sporophyte and gametophyte generations in pteridophytes provides another example; the differences in the structure and behaviour of the two generations cannot be due to the difference in ploidy (the gametophyte is haploid and the sporophyte diploid), since it is possible to obtain diploid gametophytes and haploid sporophytes by certain treatments.

How can the differentiated state be transmitted through a number of cell generations without apparent dilution? There would seem to be at least three possible situations—either (1) that in a given type of differentiated tissue the repressed state of certain genes and the active state of others is passed on from a parent nucleus to its two daughter nuclei through the replication of the DNA occurring during mitosis, (2) the genes may all be blocked during the condensation of chromatin to form chromosomes in mitosis, but selective activation or repression of the genes in the two daughter nuclei may subsequently be brought about anew by "effector" substances entering from surrounding cells, or (3) that certain cytoplasmic determinants of differentiation are transmitted from one cell generation to the next by cell lineage.

The first hypothesis postulates that the differentiated state is transmitted through the nucleus itself, the second postulates that continuity is maintained in surrounding, non-dividing cells and the third that there is transmission through the cytoplasm of the dividing cell itself.

At the present time we do not have conclusive evidence to enable us to distinguish between these three hypothesis. To distinguish between the first and second hypotheses it would be necessary to determine whether cultures from single cells still retain their characteristics, or whether biochemical differentiation is only transmitted provided the inocula used in sub-culturing contain a minimum number of cells. If transmission can occur through single cell cultures, this would support the first hypothesis, but if a minimum inoculum size is required, then this would tend to support the second hypothesis. If, on the other hand, differentiation is determined primarily by cytoplasmic factors which are transmitted to the two daughter cells during division, as suggested in the third hypothesis, then it would seem that such factors must be capable of multiplying without dilution through successive cell generations, and the idea that the cytoplasm contains self-duplicating particles, or "plasmagenes" was first suggested some years ago. At the present time we do not know which of these three alternative hypotheses is correct.

HORMONES AND GENE ACTIVITY

There are several reasons for examining the possible role of hormones in the control of gene activity. Firstly, as we have seen, all three types of growth-promoting hormone may have effects which suggest the stimulation into activity of new groups of genes. For example, the stimulation of (1) root initiation in stems by auxin, (2) flowering in long-day plants by gibberellin, and (3) bud formation in tobacco pith by cytokinin, all appear to involve the activity of genes which were previously inactive. Secondly, if hormones affect gene activity we ought to find that they increase the synthesis of certain enzymes. In the developmental effects just mentioned, we do not know what are the specific enzymes involved, but the study of animal hormones has provided a number of well-authenticated cases in which they have been found to lead to increased enzyme synthesis.

However, when we say that hormones may affect gene activity, this does not necessarily imply that hormones are directly involved in the switching on of genes. Hormones may act at one of several possible steps; for example, (1) by directly controlling gene repression, (2) by affecting the synthesis of m-RNA ("transcription" stage), (3) by affecting some step in protein synthesis ("translation" stage), or (4) by affecting enzyme activity. Thus, hormones might have an important role at some step in enzyme synthesis, without determining which genes are activated at any given time and place. Alternatively, a hormone might affect the activity of an enzyme, either directly by an "allosteric" effect (p. 277), or possibly by affecting membrane permeability, so that substrate and enzyme are brought into contact. We shall now consider what evidence there is, bearing on this problem.

Studies have been carried out on the effects of IAA on the synthesis of RNA and protein in various types of growing plant tissue, such as oat coleoptiles, and it has been found that there is frequently increased synthesis of both RNA and total protein. Fractionation of the RNA has shown that the synthesis of all fractions of transfer, ribosomal and possibly messenger, is increased by pretreatment of the tissue with IAA. Moreover, inhibitors of RNA synthesis, such as Actinomycin D, and of protein synthesis, including puromycin, have been found to inhibit growth and nullify the effect of the hormone on RNA synthesis in certain plant tissues. These observations do not, however, necessarily imply that IAA acts primarily and specifically at one stage of RNA or protein synthesis; they simply show that these latter processes are necessary for growth, but they may be only indirectly affected by IAA, through its stimulatory effect on some other processes involved in growth; the increased RNA synthesis may be the *result*, rather than the *cause*, of the increased growth. In order to gain a clear idea as to how a hormone may affect enzyme synthesis we need to study its effect on the production of a specific enzyme.

The production of α-amylase in barley endosperm (p. 130) provides a valuable system for studying the effects of a hormone on enzyme synthesis. As we have seen (p. 253), GA_3 strongly promotes the synthesis of α-amylase in isolated aleurone layers of barley. It has been shown conclusively that the increase in α-amylase activity following treatment with GA_3 involves *de novo* synthesis of the enzyme. This appears to be a relatively specific effect, since although GA_3

also stimulates the synthesis of certain other enzymes, including protease and ribonuclease, in the aleurone layer, it does not cause any detectable effect on the level of total RNA and protein synthesis in this tissue. Attempts have been made to pinpoint the precise step at which GA_3 acts on the synthesis of α-amylase, by the use of various metabolic inhibitors. It is found that under certain conditions the synthesis of α-amylase in response to GA_3 is inhibited by Actinomycin D, which is thought to inhibit the synthesis of m-RNA. Hence it is possible that GA_3 causes the synthesis of new m-RNA. However, it is not certain that Actinomycin D affects specifically the transcription of RNA, and there is evidence that it may have other effects, so that we must reserve judgement on the question of how GA_3 stimulates α-amylase synthesis.

Attempts have also been made to study the action of GA_3 on isolated cell nuclei. It is possible to isolate relatively undamaged nuclei from pea seedlings, and such isolated nuclei show the capacity for RNA synthesis when supplied with the necessary substrates. If GA_3 is added to the tissues during the isolation procedure (which involves grinding and chopping) the nuclei show greatly enhanced capacity for RNA synthesis when they are subsequently isolated. On the other hand, if the hormone is added directly to the nuclei after they have been isolated, it has no effect on the rate of RNA synthesis. Thus, it would seem that GA_3 does not act directly on the nucleus but must act first on some cytoplasmic factor, which in turn may affect the transcription process.

Studies with IAA and GA_3 have thus given no clear evidence as to how these hormones affect gene activity, but some evidence suggests that GA_3 may affect the process of gene transcription through a cytoplasmic factor. Studies on cytokinins, on the other hand, suggest that this type of hormone may be acting in a different manner.

It has been discovered that the transfer RNA for the amino acid, serine, contains $N^6(\Delta^2$-isopentenyl) adenosine (IPA), which is structurally very closely related to the natural cytokinin *zeatin* (p. 77). It has been found that this substance is very active as a cytokinin—for example, it actively stimulates cell division in tobacco pith cultures. Moreover, the *cis*-isomer of ribosyl zeatin has been found to occur in the t-RNA of several plant species, and ^{14}C-labelled benzyladenine becomes incorporated into t-RNA. On these grounds it has been suggested that natural cytokinins, such as zeatin, become attached to specific transfer RNA's, such as seryl t-RNA, and that they act in this way in RNA metabolism. It may be significant that in seryl t-RNA, IPA is attached to the RNA molecule in a position adjacent to the "anti-codon" (i.e. the part of the molecule that becomes bound to the m-RNA) and hence it may play an important role in the binding of seryl t-RNA to the complementary "codon" in the m-RNA. However, some evidence is against the view that cytokinins act as components of t-RNA. Thus an analogue of benzyladenine, 6-benzylamino-9-methyl purine, is not incorporated into RNA, due to the masking of the

9-position by a methyl group, and yet it is active in tests for cytokinins, such as the stimulation of cell division in soybean callus tissue.

The recently discovered naturally occurring growth inhibitor, abscisic acid, has been found to be very active as an inhibitor of RNA synthesis in various types of plant tissue, including leaf tissue. It also antagonizes the stimulatory effect of GA_3 on the synthesis of α-amylase in barley aleurone layer. It is not yet possible to state at what step abscisic acid is acting, but one of its first observable effects in leaf discs is a marked reduction in the content of all forms of RNA (p. 266). Moreover, ABA has been found to reduce the capacity of chromatin extracted from radish hypocotyl to support the synthesis of RNA.

Although there is no doubt that hormones have a profound effect upon gene activity, whether directly or indirectly, it is clear from the manifold hormone responses described in Chapters 4–7 that hormones have quite different effects in different tissues. For example, IAA may stimulate cell vacuolation in young fruits and the parenchymatous tissue of growing internodes, but it stimulates cell division in the cambium and it is necessary for the formation of the secondary wall in xylem differentiation. Again, GA_3 will stimulate internode extension in many plants, and the synthesis of α-amylase in barley aleurone. Thus, in plants the specificity of hormone action resides in the "target" tissue itself, and we do not find a considerable number of hormones each of which affects a rather specific target, as is the case in animals. It would appear therefore that while plant hormones regulate gene activity in many tissues, they do not determine *which* genes will be activated; that is to say, the tissue is already predetermined to respond in a particular way and the role of the hormone appears to be to stimulate the activity of genes which have been derepressed by some other mechanism. Thus, in general, the known plant growth hormones do not appear to play a primary role in determining pathways of differentiation.

There appear to be one or two notable exceptions to this generalization, however. Firstly, we have seen that auxin appears to play an essential role in the induction of vascular tissue in callus cultures and in normal differentiation (p. 158). Another well-known example is provided by the interaction between auxin and kinetin in the regeneration of buds and roots in tobacco pith (p. 156). In this latter instance gene activation does, indeed, appear to be brought about by interaction between the two hormones. However, plant development involves the sequential activation of hundreds of genes, and it is difficult to see how interaction between the small number of known hormones could specifically activate so many different genes. To meet this difficulty it might be suggested that the known growth hormones may be involved in some of the major steps in "programme selection". As we have seen (p. 22) plant development is hierarchical and involves a certain number of major pathways of differentiation (viz. differentiation into root and shoot and the differentiation of the main organs), as well as differentiation at the cellular and tissue levels. The number of such major steps in development is rather limited and possibly inter-action between the main types of hormone might play a role in controlling these steps, while differentiation at the cellular level may involve different mechan-isms, such as unequal division.

In other instances of hormonal control of development, as in bud dormancy in woody plants, and stolon development in the potato plant (p. 126), we are not dealing with an irreversible determination of the shoot apex. Thus, bud dormancy appears to involve an interaction between gibberellins and abscisic acid (p. 235), but the onset of dormancy appears to involve a rather non-specific inhibition of growth, which is overcome by winter chilling in many species. Similarly, in the interaction between IAA and GA_3 in stolon development in potato this effect only lasts so long as the hormone levels are maintained, and a switch from a stolon into a leafy shoot can rapidly be brought about by reducing the levels of IAA and GA_3, or by applying cytokinin to the stolon tip. The important feature of this latter example, as in bud dormancy, is that the action of the hormone does not appear to involve permanent gene-switching and the effects only last so long as their levels are maintained.

The alternative differentiation of axillary buds into leafy shoots or stolons in potato may depend upon the occurrence of different levels of auxins, gibberellins and cytokinins in different parts of the plant. There are several ways in which gradients in hormone levels in different parts of the plant might be effected. Firstly, auxins appear to be produced in the meristematic regions of the shoot and to be transported in a predominantly basipetal direction from the sites of synthesis. Moreover, there are auxin inactivating systems in plant tissues (p. 69) which may also give rise to gradients. Gibberellins are apparently produced both in roots and in the shoot apical region and move freely about the plant. Cytokinins appear to be formed in the roots and to be transported from there to the shoots. Thus, it is not difficult to see how the relative levels of these three different types of hormone might vary in different regions of the plant and thereby bring about differences in the developmental patterns of shoot apices.

In spite of the instances cited above in which hormones appear to stimulate differentiation, most of the effects of hormones appear not to be concerned with such gene activation but with the stimulation of activity in genes which are already in a derepressed state, but which are not able to effect enzyme synthesis for some other reason. Indeed, the majority of hormone effects appear to be attributable to the stimulation of growth in organs which are already deter-mined (e.g. internodes or fruits). Again, within the plant as a whole, hormones appear to act in the correlation of growth in different regions of the plant. The possible manner in which they act in correlation has already been discussed in Chapter 6.

THE CONTROL OF FORM

So far we have discussed some of the possible control mechanisms of differentiation at the molecular and cellular levels, but we have still not considered the problem of the overall determination of form in the whole organism, or in an organ such as a leaf or a flower. Clearly, the attainment of the form and structure of the fully developed leaf is a highly complex process and calls for a high degree of co-ordination of cell division and growth. As we have seen (p. 54), relatively simple differences in form, such as occur with different shapes of fruit in gourds, can be explained in terms of differences in rates of cell division along certain axes. However, we cannot yet say what determines the planes of cell division and differential rates of growth in various directions. When we come to more complex forms such as we see in a stamen or a carpel, the problem of co-ordination of cell division and growth becomes much more difficult to understand and at present we have no idea how this is achieved. It is clear that nuclear genes are involved, since many differences in form, including those affecting fruit shape of gourds, are inherited in a simple Mendelian manner. Nevertheless, it is equally clear that cytoplasmic factors also play a vital role, since the planes of cell division appear to be determined by the polarity within the tissue mass as a whole.

It must be admitted that we have not even begun to understand how one of the most conspicuous attributes of living organisms, their characteristic form, is determined, and morphogenesis offers one of the most challenging unsolved biological problems.

NON-GENIC FACTORS IN DEVELOPMENT

In any discussion of the control of development it is axiomatic that in the final analysis all aspects of development are determined by the genetic make-up of the species, interacting with the external factors of the environment. In speaking of the "genetic make-up" of a plant one normally refers to the information coded in the DNA of the nucleus. However, we cannot assume that the total genetic constitution is entirely attributable to the DNA of the nucleus. Indeed, a considerable number of instances of "cytoplasmic inheritance" (in which certain characters of the offspring are transmitted through the maternal cytoplasm) are known for plants, and we now know that certain cytoplasmic organelles, notably the plastids and mitochondria, have some degree of autonomy and indeed contain their own DNA.

Thus, we cannot exclude the possibility that certain aspects of development may involve other factors, besides those controlling gene activity. Such non-genic factors probably include certain "physical" (i.e. non-physiological) properties of cells and tissues. We have already seen the importance of *external* environmental factors, such as light and temperature, in plant development, but we must also consider the effects of other physical factors, such as surface tension and diffusion gradients of oxygen, arising and operating within the cells and tissues themselves. For example, attempts have been made to account for the position of cell walls in terms of physical laws. Earlier workers drew a parallel between the shapes and arrangements of cells in a tissue and those of groups of soap bubbles. Now, the shapes of soap bubbles can be interpreted in terms of the effects of surface tension, which tends to make them adopt forms with the least possible surface area. Errera suggested that the shapes of cells can be similarly interpreted, although it is questionable how far cell walls can be regarded as equivalent to the almost weightless films of soap bubbles. Moreover, many cell types, such as cambium cells, do not appear to adopt a form with a minimal surface area. It is possible, however, that Errera's laws are applicable to isolated cells, or groups of cells, which will be free of the pressures and other influences from surrounding cells which occur in a tissue mass. Thus, the position of the walls formed during the early development of plant embryos may follow Errera's law and be determined by surface tension effects.

Although we cannot elaborate further on this theme here, it is clear that we should avoid the danger of assuming that all aspects of growth and development can be accounted for in terms of the information encoded in the DNA of the nucleus.

FURTHER READING

General

Various articles in *The Molecular Basis of Life* (Readings from *Scientific American*), W. H. Freeman & Co., San Francisco and London, 1968.

BARRY, J. M. *Genes and the Chemical Control of Living Cells*, Prentice-Hall Inc., Englewood Cliffs, N.J., 1964.

BONNER, J. *The Molecular Biology of Development.* Clarendon Press, Oxford, 1965.

LAETSCH W. M. and R. E. CLELAND. *Papers on Plant Growth and Development*, Little, Brown & Co., Boston, 1967.

STEWARD, F. C. *Growth and Organization in Plants*, Addison-Wesley, Reading, Mass., 1968.

TORREY, J. G. *Development in Flowering Plants*, MacMillan and London, 1967.

WADDINGTON, C. H. *Principles of Development and Differentiation*, MacMillan, New York and London, 1966.

WATSON, J. D. *Molecular Biology of the Gene*, W. A. Benjamin Inc., New York, 1965.

More Advanced Reading

BRINK, R. A. Phase change in higher plants and somatic cell heredity. *Quart. Rev. Biol.* 1962.

GIBOR, A. *Acetabularia*—a useful giant cell. *Sci. Amer.* **215,** 118, 1966.

HESLOP-HARRISON, J. Differentiation. *Ann. Rev. Plant Physiol.* **18,** 325, 1967.

JACOB F. and J. MONOD. Chapter in *Cytodifferentiation and Macromolecular Synthesis*, Academic Press, New York and London, 1963.

KLEIN, R. M. The physiology of bacterial tumours in plants and of habituation. *Encycl. Plant Physiol.* **15**(2), 209, 1965.

TREWAVAS, A. Relationship between plant growth hormones and nucleic acid metabolism, in *Progress in Phytochemistry* (Ed. L. Reinhold and Y. Liwschitz), Interscience Publishers, London and New York, 1968.

Index

295